Business Week Guide to Multimedia Presentations

Robert L. Lindstrom

Osborne **McGraw-Hill**

Berkeley New York St. Louis San Francisco
Auckland Bogotá Hamburg London Madrid Mexico City
Milan Montreal New Delhi Panama City
Paris São Paulo Singapore Sydney Tokyo Toronto

Osborne **McGraw-Hill**
2600 Tenth Street
Berkeley, California 94710
U.S.A.

For information on translations or book distributors outside of the U.S.A., please write to Osborne **McGraw-Hill** at the above address.

Business Week Guide to Multimedia Presentations

1234567890 DOC 9987654

ISBN 0-07-882057-X

Publisher
Lawrence Levitsky

Acquisitions Editor
Scott Rogers

Project Editor
Bob Myren

Technical Editors
Gordon G. Miller III
Sharon P. Pitt

Computer Designer
Peter F. Hancik

Illustrator
Marla Shelasky

Series Design
Kris Peterson

Quality Control Specialist
Joe Scuderi

Cover Design
Clement Mok Designs, Inc.

This book is dedicated to Karen, Katie and Whitman, a family of unfathomable patience and impeccable credentials as human beings.

This book is also dedicated to the memory of my good friend Nick Walker who made the summit ahead of the rest of us.

Table of Contents

1

THE MULTIMEDIA INCENTIVE I

2

MULTIMEDIA AND THE ART OF
PERSUASION 23

3

THE MULTIMEDIA TOOLBOX:
AN OVERVIEW 47

—4—

DESTINATION MULTIMEDIA: PLANNING THE PRESENTATION 89

5

PRESENTATION CREATION: WORKING WITH AUTHORING SOFTWARE

6

THE ELECTRONIC CONVERSATION: UNDERSTANDING INTERACTIVITY

7

SENSUAL PERSUASION: UNDERSTANDING MEDIA AESTHETICS 195

8

PICTURE THIS: WORKING WITH GRAPHICS 231

—9—

THE AURAL DIMENSION: WORKING WITH SOUND **279**

12

MULTIMEDIA 911: FINDING PRODUCTION ASSISTANCE 411

Acknowledgments

This book was not a solo climb.

The project would not have coalesced without the vision of Larry Levitsky, the expedition leader at Osborne/McGraw-Hill.

The logistics and planning were guided by Scott Rogers. Enthusiastic and optimistic even when the ridge seemed to be crumbling, Scott kept the pages turning and the ropes untangled.

Bob Myren and his editorial team, most notably Kathryn Hashimoto, exhibited uncommon fortitude under pressure and are responsible for the polish and consistency of the final product.

My sincere gratitude and respect goes out to technical editors Gordon G. Miller III and Sharon P. Pitt for their efforts to keep the book accurate and on track and for making complex information accessible to non-technical readers. Thanks also to Gordon for his encouraging words on the Net, and for his patience as DHL, Fedex, and I chased him around the country with draft pages to read while he was busy doing his job as director of the Multimedia Lab at Virginia Tech.

Contributing Editors

Special acknowledgment goes to the contributing editors, whose expertise in a wide range of disciplines added both depth and scope to the project:

Mike Agron is director of the New Media Presentations division of Decker Communications in San Francisco. The division is charged to help clients improve communication by integrating multimedia technology with presentation skills and content development.

Bert Decker is chairman of Decker Communications Inc., a San Francisco-based communications consulting and skills training organization. He is the author of *You've Got to Be Believed to Be Heard* (St. Martins Press), a book based on the premise that successful communication depends on gaining the emotional trust of others.

Glenn Johnson is multimedia communications director for Graphix Zone in Irvine, CA, where he is responsible for producing and delivering multimedia seminars, training experiences and marketing collateral using both the Macintosh and Windows platforms.

Nicole Lazzaro heads ONYX Productions, an interface design and multimedia production company based in Oakland, CA. She is co-founder of the San Francisco State University Multimedia Studies Program and teaches classes in interface design.

Gordon G. Miller III, technical co-editor, is director of the Multimedia Lab at Virginia Tech, where he coordinates and directs a multimedia research and testing initiative. His areas of research include interface design, media integration, cross-platform development, and digital video development.

LaTresa Pearson is associate editor for *Presentations* magazine, which focuses on presentation technology and techniques. LaTresa is a specialist in interactive multimedia technology and applications for both the corporate and consumer markets.

Sharon P. Pitt, technical co-editor, is a multimedia specialist with Educational Technologies at Virginia Tech. Her areas of expertise include graphic design, imaging, 3-D modeling, and digital video production, as well as user interface issues and human factors.

▶ **Peter M. Ridge** is engineering manager at Creative Labs, Inc. and co-author of *Sound Blaster: The Official Book* (Osborne/McGraw-Hill). His hobbies include skiing, tinkering with MIDI music and playing PC games.

▶ **Jason Roberts** is the founder of Panmedia, a New York City-based multimedia production studio, and author of *The Macintosh Hard Disk Companion* (Sybex), and *The Complete Director* (Peachpit). His articles have appeared in *The Village Voice*, the *San Francisco Chronicle* and *Metro* magazine.

▶ **Michael Utvich** is a recognized expert in computer media, publishing and automation. His consulting work includes the areas of high technology training, marketing and system implementation. He is the author of five books on computer media.

I am reminded as I send the final files of manuscript to the publisher of what Frank Lloyd Wright once said about his profession: "A doctor can bury his mistakes but an architect can only advise his client to plant vines."

The same principle applies to writing a book. Once I finish design and construction, it is there for all to see. I offer it with the hope that it will be of good use to those who read it, and with the hope that there will be no need to plant vines.

Introduction

> According to the World Health Organization, sexual intercourse among human beings occurs more than 100 million times daily.

Your first thought is probably, "What does that statistic have to do with a book about multimedia business presentations?" The answer is, absolutely nothing. It's just a cheap, hackneyed way of grabbing your attention. But it also serves as an object lesson crucial to successful presentation: first and foremost, you must command the attention of your audience.

Now that I have your attention, consider *this* fact:

> According to the U.S. Census Bureau, from 1960 to 1989 more than 91 percent of U.S. households made at least one move from one home to another. Nearly 50 percent of the population made a move in the period from 1985 to 1989.

While the first statistic might stand on its own as an attention getter, the second needs serious help. In a traditional slideshow the obvious solution would be to create a chart to express the figures visually. The message could be emphasized by superimposing the chart on an illustration of a moving van, or perhaps over an image of a family packing their belongings.

But now imagine the same information accompanied by a video of a family busily loading a rental truck. The audience sees an animated graph grow vertically and horizontally to depict the trend across time. As it does, the video divides into two smaller videos, then four, then eight and so on. The music track swells in intensity, hitting a crescendo as what are now hundreds of tiny videos of people packing and moving fill a 3-D outline of the United States. Now imagine that by touching any area on the map, the screen instantly displays the specific migration figures for that region.

Assuming it was the goal of the presentation to impress upon the audience the tremendous mobility of our society, which of the two approaches would be most effective?

In a nutshell, that is what this book is about—conveying information and making a point in the most persuasive way possible. It is both about the process of presentation, and about the astonishing new technologies that exist to make the process more efficient and more effective.

Profiting from Presentation

There is much to be gained from the successful linking of presentation skills and technology tools. You might, for instance, find a faster route to higher margins, a less costly way to produce support materials, or a better way to formulate and express your ideas. This book will address all of those benefits and more. But it should be understood from the outset that the central objective of this book is *profit*—both the above-the-line variety that comes from acquisition of knowledge and the bottom-line type of profit that means more money in your pocket.

Techno-dependency

The Business Week Guide to Multimedia Presentations is written from the premise that business and the technology used to conduct business have become joined at the hip. Without the communications infrastructure, without computers and computer networks, without office productivity devices, commerce as we know it would come to an abrupt halt. Business has become technology-dependent. Business careers have become technology-dependent, and the effective use of technology has become a business skill in its own right.

Emphasizing Soft Skills

The effective and productive use of business communications technology goes far beyond selecting products and configuring systems. The most elaborate communications system ever invented is still only as effective as the people using it. Communication *soft skills*—what you say and how you say it—have always been and continue to be the most important success factors in business.

This book discusses the fit between the tools for multimedia production and the basic principles of presentation with this simple prediction in mind: If you are not already presenting with multimedia technology, you will be, sooner or later.

Who Needs This Book?

The Business Week Guide to Multimedia Presentations is based on the assumption that you bought or borrowed it for one or a combination of the following reasons:

▶ To learn how to harness the power of integrated digital media for presentations

▶ To learn how to become a better business communicator

▶ To learn how to improve your image and advance your career

▶ To learn how to be more productive in your work

▶ To keep the door propped open

This book was written for anyone who is interested in multimedia as a presentation tool. It is for those who are already advocates of the technology, but wish to learn more about what it can do for them and their organization. It is for skeptics who need more convincing. It is for occasional presenters who make only a handful of mission-critical presentations each year. It is also for power presenters—those individuals who wield a laser pointer like Luke Skywalker handles a light sword.

How Is This Book Organized?

For the digitally averse, the approach to the components and capabilities of multimedia are explained one step at a time. There is enough information here to set you on your way, though not so much technical detail that it will crash your mental processor. Each chapter is organized so that the technical details, for the most part, increase as the chapter or subsection progresses. If you find yourself bogging down in the technical basics, skip ahead. Come back to the sections as you feel comfortable or need them for reference.

If the technical information seems remedial to you, focus instead on the cross-pollination issues of presentation techniques and presentation technology.

You will find the subject of multimedia presentation addressed in five areas: Concepts, strategies, technology, tutorials, and case studies. How you actually use the book is largely a matter of the perspective you bring to it:

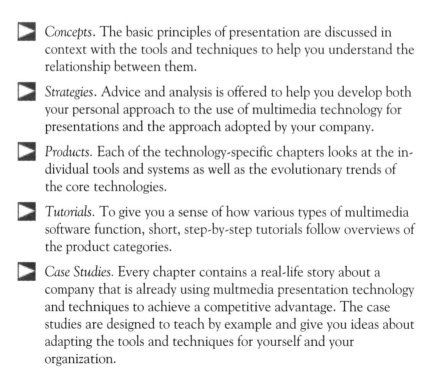

Concepts. The basic principles of presentation are discussed in context with the tools and techniques to help you understand the relationship between them.

Strategies. Advice and analysis is offered to help you develop both your personal approach to the use of multimedia technology for presentations and the approach adopted by your company.

Products. Each of the technology-specific chapters looks at the individual tools and systems as well as the evolutionary trends of the core technologies.

Tutorials. To give you a sense of how various types of multimedia software function, short, step-by-step tutorials follow overviews of the product categories.

Case Studies. Every chapter contains a real-life story about a company that is already using multmedia presentation technology and techniques to achieve a competitive advantage. The case studies are designed to teach by example and give you ideas about adapting the tools and techniques for yourself and your organization.

How Should You Read This Book?

There are several ways you can approach the subject matter. I advise you to take some time to consider a reading strategy:

> If you prefer to plow through page by page, chapter by chapter, that's your privilege. Chapters 1 and 2 are designed to give you some background on multimedia presentations and to put the technology into context before beginning a discussion of the actual products and process.

> If you know enough about multimedia technology that you are ready to begin developing presentations, but will be contracting out for multimedia development services, read the last chapter first. It will alert you right away to what is available and how to go about dealing with freelancers and outside services. But, before you start making calls and taking bids, read Chapter 4 about designing and planning a multimedia presentation.

> If you plan to author your own presentations and are in a hurry to get started, concentrate on Chapter 5, which addresses authoring, then move straight to Chapter 9, working with sound. Sound is an easy and accessible place to begin. Using the software available on the CD-ROM you can begin to see results almost immediately.

> If you want a quick overview of what is encompassed by the multimedia toolset, Chapter 3 will start you off with a brief but comprehensive overview of the product categories and how they interrelate. Go from there to the more specific technical chapters.

> If you are most interested in how graphics, sound, and motion work to enhance the presentation process, you may want to head directly for Chapter 7, which looks at the conscious and subconscious power of media aesthetics.

This book is not intended to be the definitive work on multimedia technology or its application. Likewise, it is not meant as a complete manual on the art of presentation. Rather it looks at how the two areas, technology and presentation principles, commingle to provide an opportunity for improved business communication.

In my opinion, how multimedia technology can and is being used to achieve dramatic improvements in business presentation is, at heart, far more important than how the technology operates. But the fact is you must have a firm grasp on the latter to effectively pursue the former.

As I frequently tell people, there are four simple rules of good multimedia presentation:

Rule #1: Have something to say.
Rule #2: Say it clearly.
Rule #3: Say it like you mean it.
Rule #4: No amount of technology can help you with Rule #1.

Robert L. Lindstrom 7/94

If you find this book to your liking, turn to the coupons following the index for an offer to receive a **FREE ISSUE** *of the* **Multimedia Presentations Report,** *Robert L. Lindstrom's monthly newsletter that will help you stay on top of the trends, the tools, and the resources.*

About the CD-ROM

The CD-ROM that comes in the back sleeve of this book is loaded with software. If you have a PC or Macintosh computer equipped with a CD-ROM drive, you can access free clip media and software demos, and test-drive versions of many of the software packages mentioned in the book. What's more, the software can be instantly downloaded from the CD upon purchase!

There is at least one example of each major category of multimedia software—enough firepower to get you started right away if that is your intent. The tutorials in the book can be read as a basic introduction to multimedia authoring, media creation, and editing, or you can use the tutorials to test the actual software.

For no charge you can load the programs, work with them, experiment, and explore. Rarely will you have an opportunity to experience all of these packages on your own before buying them. Few dealers have presentation software up and running for demonstration; none will have the breadth of multimedia presentation programs you'll find here. Once you find the tools you are looking for, the software can be purchased and installed instantly by calling InfoNow toll free at 1-800-640-1853.

The disc also contains a rich selection of clip media. The images and sounds are samples from some of the best clip media companies in the industry. Use the clips as you experiment with the software or work with the tutorials.

The disc includes:

 Free sample media clips of still photos, music, sound effects, animation and video.

▶ Ordering information for receiving the complete clip media collections.

▶ Test drives of some of the most popular PC software applications for developing multimedia presentations.

▶ Complete, installable versions of the PC programs that you can purchase by calling the listed toll free number to receive the necessary unlocking code.

▶ For Macintosh users, self-running demos and some working models of many of the products mentioned in the book and used as tutorials. (A disc with fully functional versions of the Macintosh products is available by mail to purchasers of the book. To receive this disc, call InfoNow at 1-800-640-1853 between the hours of 7 A.M. and 9 P.M., Mountain Standard Time)

▶ The Osborne/McGraw-Hill publication catalog.

PC Users

System Requirements for PC Users

An IBM-PC–compatible computer with a CD-ROM drive, Windows 3.1 or higher. A sound card is recommended. You may need significant amounts of hard disk space to test-drive or purchase some of the programs on the CD-ROM.

Accessing the Test Drives and Installable Programs

To run the programs that are included on the disc as fully functional test drives, first insert the CD into your CD-ROM player and run the install program.

From the Windows Program Manager select the Run option. Type the drive designation for your CD-ROM player and type **setup**.

The installation program will install both the Osborne/McGraw-Hill catalog and the InfoNow Software Catalog.

Answer the prompts on the screen as they appear. You will be asked if you want to modify your AUTOEXEC.BAT file. This is necessary for running the test drives. After you have made your selections you will need to reboot your system in order for the test-drive function to work.

After the reboot, go to the newly-created Catalog group in the Program Manager, and double-click on the Start Here icon. This will start the program. Now you can run the InfoNow Catalog by clicking on the Software button from the main screen.

To run the animated help system in the InfoNow catalog select Help from the menu and then the Run Animation option. The animation describes the entire catalog and how to test-drive, demo and buy/install the software programs.

You can browse the available software programs by:

▶ Clicking on the Prev. and Next buttons at the bottom of the screen. The screen will give you a description of the product and the buying information.

▶ Selecting the Sections box.

▶ Selecting the Menu Browse option for a title search.

The features box lets you scroll down to learn more about each software program's features, benefits, and system requirements.

Going for a Test Drive

To test-drive a software package select the Test Drive icon. You will be asked to call the toll free number on the screen for an unlock code. This is for security reasons and will take just a few minutes. Call InfoNow at 1-800-640-1853 between 7 A.M. and 9 P.M. Mountain Standard Time. Follow the instructions on screen for entering the unlock code. You will be able to use the full-featured test drive five times, for as long as you wish at each sitting, before the security system renders the test-drive function inoperable.

note

If you have any technical questions with the CD-ROM, call InfoNow at 1-800-640-1853 and they will help you resolve your problems.

Running a Demo

The self-running demonstrations can be started by simple clicking on the Demo icon. When the program concludes it will return you to the InfoNow catalog menu.

Placing a Software Order

To purchase the software and make it instantly available for installation select the Order icon. You will be prompted to call the toll free number. After supplying your payment information you will be issued a special unlock code. Follow the instructions on the screen to unlock the software you have purchased. You can begin using it immediately.

In a few days you will receive the entire software package by mail, including all manuals and backup disks.

Accessing the Free Media Clips

There are three ways to access the media clips after you have inserted the disc:

▶ Browse through the InfoNow Software catalog for full information about each set. The browser will tell you how to find the media clip you would like to preview.

▶ Locate the media clips by filename with your Windows File Manager or other such utility and transfer them directly to your hard drive.

▶ If you are running the appropriate software, such as a multimedia presentation authoring program or a drawing package, you can load the files directly from the CD into the application.

Mac Users

Accessing the Self-running Demos and/or Working Models

The disc does not contain the full working Macintosh versions of the software. You will be able to run demos for some of the packages and work with partially enabled versions of others. Information on screen will tell you how you can order the Macintosh products or receive the Macintosh version of the InfoNow catalog by mail.

note

A disc with test-drive versions of these programs is available immediately from InfoNow. Call 1-800-640-1853 and InfoNow will mail you the free disc.

Caution: Some of the demos require 8MB of RAM. If your system has less RAM you may not be able run some demos.

To run the CD-ROM on your Macintosh, insert it into the player and click on the CD icon when it appears on the screen. You will have the option of selecting either the Osborne/McGraw-Hill catalog or the InfoNow catalog.

If you choose the Osborne/McGraw-Hill catalog, you can access the demos from within it by selecting the Software button.

From the InfoNow catalog you can browse the available software programs by:

▶ Clicking on the Prev. and Next buttons at the top of the screen. The screen will give you a description of the product and the buying information.

▶ Selecting the Contents box.

The features box lets you scroll down to learn more about each software program's features, benefits, and system requirements.

Running a Demo

Start a demonstration of the software by clicking on the Demo icon.

Accessing a Working Model

Select the Working Model icon to launch a version of the software that has been partially disabled. You will be able to use most of the functions but will not be able to save your work.

Placing a Software Order

If you are interested in buying the software, select the Order icon. You will be given a toll-free number to call. By supplying payment information over the phone you will receive the complete software package in a few days. You can also order, at no charge, the CD-ROM that includes fully functional Macintosh software applications not found on the CD-ROM that comes with the book.

Accessing the Free Media Clips

There are three ways to access the media clips after you have inserted the disc:

▶ Browse through the InfoNow Software catalog for full information about each set. The Features box will tell you how to find the media clip you would like to preview.

▶ Locate the media clips by filename and transfer them directly to your hard drive.

▶ If you are running the appropriate software, such as a multimedia presentation authoring program or a drawing package, you can load the files directly from the CD into the application.

Good communication is as stimulating as black coffee, and just as hard to sleep after.
—Anne Morrow Lindbergh

THE MULTIMEDIA INCENTIVE

With the arrival of digital multimedia, the personal computer has crossed the threshold from the world of the static to the realm of the dynamic, where information is experienced with the senses and emotions as well as the intellect. The new tools, techniques, and systems have the power to dramatically extend the reach and the impact of business communications for individuals and organizations.

Digital Presentation

Multimedia is not new to presentations. Presenters have been telling their stories and selling their messages with multiple media types (text, graphics, sound, animation, and video) for decades. In fact, just about every media recording or playback device ever invented has at one time or another been adapted for business presentations. The birth of visual aids undoubtedly goes back even further, to the first time a protohuman drew a picture in the dirt or on a cave wall to clarify a concept for his lowbrow brethren.

Early religious art or primitive presentation?

But presentation tools and techniques that might have worked well in the dim light of prehistory would be lost in the glare of the age of information, an age in which we mass-produce information on an assembly line of computers, networks, satellites, and cables. The "new" in what is called multimedia, or sometimes new media, refers to the remarkable advances in technology that allow text, graphics, audio, and video to be integrated and adapted in digital form. Media types that once had distinct methods of creation, storage, transport, and access have found a common language in the digital 0s and 1s, the bits and bytes, of computers.

Just as advances in computer technology made it possible years ago to translate numbers, then text, then graphics into digital information, now both audio and video have joined the digital world as well. What that means for presentations and communications of all types is most evident in the example of audio compact discs, one of the most successful consumer electronics products in history.

CD-audio stores sound information as mathematical data. When played back, the music is reproduced exactly as it was recorded. You hear none of the loss of quality or added noise associated with vinyl records or magnetic tape. In addition, the music tracks on CD are randomly accessible. Any track can be played in any order by programming a CD player. With the right equipment, you can select and play not only any track but any single note or combination of notes—a feature that would be considered excessive by most consumers.

Digitization and the programming power of computers free audio and video from the constraints of tape and film. Sounds and images can be accessed in any order, rearranged, edited, and manipulated as desired. In digital form, media can be reproduced with extraordinary accuracy, beamed through the air or sent by wire. It can be shaped, enhanced, reused, altered, and integrated with other digital information in infinite combinations.

The technology that makes digital media possible will be discussed in detail in later chapters specific to each medium. At this point, before looking at multimedia as a technological development, it is important to understand multimedia as a communications concept. Multimedia represents a new way of thinking about information. The "digital revolution" is as much about the revolution in communication models—business, education, entertainment, personal—as it is about revolutionary technologies.

The changes wrought by a world gone digital have broad and far-reaching social implications. The macrocosmic trends in communications models and technology will affect every level of business operation, but the concern here is that you discover the immediate and evolving benefits of multimedia as a presentation tool and strategy. Before you can take advantage of it, you must first know what multimedia presentation can do for you: how can it give you and your organization a competitive edge in the information age?

The Info Avalanche

The most striking paradox of the information age is this: the more information we produce, the less time we have to assimilate it. Even so, we are told to expect the volume of information to continue to multiply exponentially, theoretically without end. That presents a profound challenge to businesses and business people, who must find ways not only to distill information into knowledge (see Figure 1-1) but also to overcome the information repellents that people naturally exude as the data swarms increase.

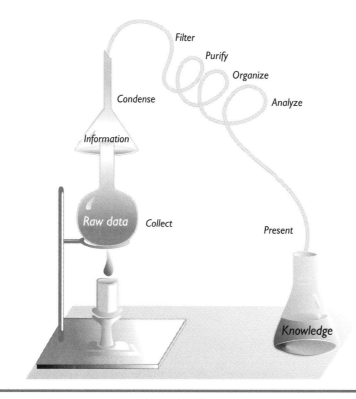

Figure 1-1   Multimedia assists with the distillation of data

In an environment where an increasing volume of information vies for the attention of individuals and businesses with a finite capacity to absorb what is being offered, your message can easily be lost, misinterpreted, or ignored outright. Business communication must happen faster; it must be precisely targeted, and it must hit with maximum impact. How well you execute those three requirements will determine, to a large extent, how you are measured by your customers and your competition.

Multimedia Attributes

Multimedia directly addresses the problems of information overload and overtaxed attention spans. It confronts bottom-line considerations of productivity and sales revenues even as it offers above-the-line solutions for enhanced communication, learning, and decision-making.

The hallmarks of effective multimedia are

- Multisensory engagement
- Dynamic, time-based information
- Feedback and interaction
- Customization and targeting
- Flexibility and changeability
- Creativity and experimentation

Keep in mind from the beginning of your exploration into multimedia that its power does not stem from the addition of one medium or another. Its power is *synergistic*. Just as it takes many instruments playing in concert to produce a symphony, the artful combination of media elements characterizes multimedia as a new and advantageous communications concept.

Within that context, you as the presenter or the creator of a multimedia presentation become the conductor and composer. You have at your disposal a complete range of instruments. How you choose to arrange them and where you put the emphasis is up to you. Again, like the conductor or composer, you must train yourself to understand the principles and techniques of integrating media for better communication.

Multiple Levels of Engagement

Computers cannot be sincere. Nor can they be emphatic, enthusiastic, or empathetic. They can only collect, process, and store information. The power of information is not contained in the data, but in the ideas, the emotions, and the actions it triggers in the people who encounter it. Information is alive, evolving, and ever-changing.

Digital technology has made it possible to collect, process, analyze, and store information with astounding speed and accuracy. Yet, until relatively recently, it has done little to assist business people with the equally challenging task of communicating the dynamic nature of information in human terms.

Humans are multimedia communicators. We experience our world through our senses (see Chapter 7) and express our needs and desires with a complex variety of verbal and visual signals. A study by Dr. Albert Mehrabian, a specialist in interpersonal communications at the University of California, found that words alone (verbal) account for 7 percent of the impact of face to face communications. Vocal communication, associated with characteristics of the voice such as tone and inflection, accounts for 38 percent. Visual communication, including facial expressions and gestures, accounts for 55 percent of the total impact. (See Figure 1-2.)

Monomedia communication, in which we take advantage of only one media at a time to convey our message to others, limits the communication process by ignoring or neglecting the other levels on which human beings naturally send and assimilate information:

▷ *Written language,* when skillfully employed, is an exceptionally powerful communications medium, but the finest speech writing cannot fully replace the speaker's physical gestures, vocal inflections, or facial expressions. Consider the difference between a printed résumé and a personal interview, or the difference between a written quotation and the same thought coming directly from the person quoted.

▷ *Illustrations* and *photographs* can tell stories, grab attention, and describe people, places, or things. But even the most finely detailed im-

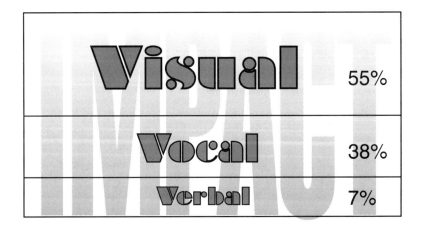

Figure 1-2   Communication dynamics

age is frozen in time and locked in silence. A photo of a busy assembly line might help an audience better understand the subject, but it does not convey the flurry of activity or the hum of the machinery.

▶ A *chart, graph,* or *diagram* adds a spatial dimension to printed data but can only suggest the element of time using dates and scales. A chart that grows, expands, changes color, and comes together or flies apart tells an audience what the figures imply and not just what they represent in concrete terms.

The missing elements in the preceding list are motion and sound. The addition of one or both (multimedia) engages the senses on multiple levels. They bring an audience closer to the information, allowing the audience to experience the information in a way that is much closer to the multisensory experience of everyday life.

Information in Time

Elements of presentation that do not convey movement and do not produce sound are said to be *static.* In a sense, the information is frozen in time. Our eyes move, but the information does not. A book, a document, a painting, or a photograph are examples of static media. We examine information in static form as we would a dinosaur bone, turning it over in our hands for a minute or a day, considering what it has to tell us.

Yet we live in a dynamic world. Time defines our perceptions. It puts information into context and it establishes relationships. Using time-based media (motion and sound), it is possible to present the visual unfolding of information, showing change over time. We can see changes happening with our own eyes that static media can only describe to us (see Figure 1-3). Dynamic media presentations demonstrate the process of change using movement and sound. The addition of pace and tempo keeps the information in context with our experience.

The dynamic media of motion and sound produce information events. In an *information event,* the audience is party to the unfolding of information in real time. They see and hear it happening. They experience the emotions it evokes. As mentioned, the overall impact is a synergistic combination of all the information channels that are being used. Of course, multimedia presentation still takes advantage of static media. But it extends the communication power of text, illustration, and

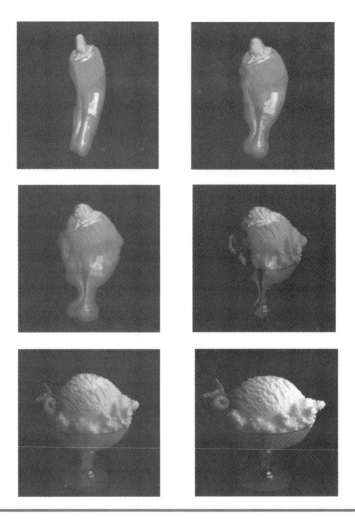

Figure 1-3   Dynamic morphing effects, here created by HSC
Digital Morph, depict transformation over time

photographs by adding the dimension of time and making them part of the information experience.

Presenter/Audience Interaction

In addition to being creatures of time and of the senses, we are social creatures. We communicate to live. Human communication is a complex process, but its key characteristic is that it works in two directions, often simultaneously. We go in search of the information we need and we respond to requests from others for the information they require. The model is that of a conversation (see Chapter 6).

Every good presentation offers the opportunity for interaction; the audience responds to the presenter, and the presenter responds to feedback from the audience. Experienced presenters learn to sense audience response even without verbal comments or questions. "Chalk talks," flip chart presentations, and overhead projector presentations are conducive to presenter/audience interaction, but visual support is limited to what can be drawn by the presenter or the small set of prepared visuals. Certain presentation technology, 35mm slides for example, tends to lock the presenter into a prearranged sequence of information, forcing the presentation to become more monologue than dialogue, more lecture than conversation.

In contrast, information stored in a digital format can be accessed randomly, a feature that comes into play both when creating and delivering a multimedia presentation. The computer program responds to the user on the fly. Because content is digital, it can all be stored in one place; and because computers can store vast amounts of data or access it over networks from other computers, a presentation can theoretically have access to a bottomless well of dynamic information. As in a conversation, information is both delivered and received. Information coming in can be analyzed and supplemented before being output, improving the odds that the right information is delivered to the right people at the right time (see Figure 1-4).

User-Driven Presentation

Building interactivity into a multimedia presentation not only changes the nature of information exchange, it changes the notion of who is the audience and who is the presenter. In-house and public-venue kiosks that use multimedia to deliver and collect information are audience-driven presentations. The presenter, in absentia, controls the information that is accessible to the user/audience and, to some extent, controls the way the user navigates the presentation, but the user does the rest.

Kiosks are popping up like mushrooms after a rain in stores, malls, government buildings, and corporate lobbies, as organizations look to

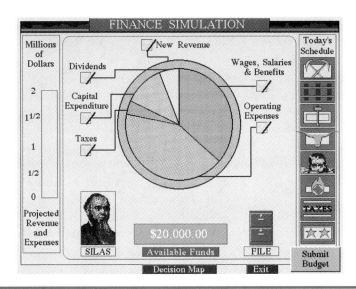

Figure 1-4   One example of interactive presentation is a live financial simulation, here created using Allen Communication's Quest, in which figures are entered and charts are generated on the fly

multimedia technology to present more information to more people in a manner that is convenient to the audience. Kiosks are also becoming an increasingly common sight at trade shows, where they take the sales pressure off of prospective customers by allowing them to seek out the information specific to their interests and at their own pace. (See the case studies in Chapters 4, 7, and 11.)

These sorts of user-driven presentations, which are closely related to and sometimes indistinguishable from computer-based multimedia training programs, let users access information not only at their particular level of interest and their own pace, but also at their own level of comprehension. A few companies have even gone so far as to create sales presentations in which the customer or prospect enters information into the program and receives real-time feedback on pricing and features (see the case study later in this chapter).

Because a presenter is not there to hold the attention of an audience, user-driven presentations must be stimulating and engrossing. They also

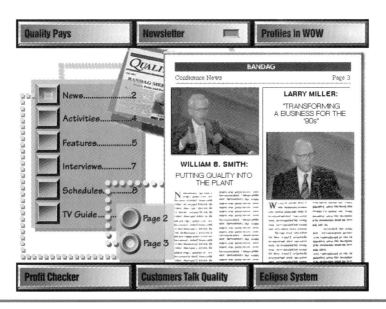

Figure 1-5   Interactive multimedia newsletters are essentially presentations without the presenter. ©1994 Iconic Interactive, Inc., San Francisco, CA

must communicate the subtextual and emotional aspects of the information, a job normally left to the presenter. Multimedia techniques are ideal for grabbing the audience and holding their attention with color, motion, sound, and interactivity (see Figure 1-5). Nuances of emotion can be added to otherwise static and lifeless material, thereby heightening the experience for the audience. In a sense, multimedia lets the audience see the expression on your face even when you are not there. Or, as one multimedia enthusiast expresses it, "It's like being there without being there."

Custom-Tailored Communication

It is a fact of human nature that everyone appreciates being addressed by name. During any communication experience, everyone wants—and deserves—to be treated as if they are the most important people on the planet at that moment. A classic example of how marketers try to take advantage of this natural human trait can be seen in the direct mail and magazine ad campaigns that include (sometimes clumsily) personalized letters that address the occupant by name.

Following the same principle, multimedia presentations are customizable, not only with client names and specific data, but with photographs, videos, and even specially selected music. Graphics software makes it possible to alter portions of images to make them unique for each audience, without the need to recreate the image in its entirety (see the case study in Chapter 8).

Some multimedia presenters already make a habit of taking a camcorder or digital still camera on the road as part of their multimedia presentation gear. Before the presentation, they shoot footage that has personal relevance to the client, customer, prospect, or audience: offices, logos, products, individuals, and so on. With the proper equipment, the footage is digitized and incorporated into the presentation at preselected points.

Customized presentations are nothing new, but the ease with which a core presentation can be fitted to different audiences without the need to completely reauthor the audio and visuals makes the process far more productive and affordable than it has ever been before.

Recycling and Repurposing Content

In the same way that digital multimedia allows presentations to be customized to suit the needs of a particular audience, the flexibility of the digital data types gives new life to old material and keeps new material from going stale. Once converted to digital format, snippets of video, old photographs, and past presentations can be given fresh graphics treatments or updated with special effects. A defunct video presentation, for example, that would be too costly to reedit and update with titles, a sound track, or graphics might find new life as part of an interactive multimedia presentation.

Repurposing, the adaptation of courseware or other materials to new uses, is already common in educational and training applications. Applied to presentations, repurposing of materials from other departments, even other businesses, offers a means of acquiring content without incurring the expense or time of creating it from scratch. In many cases, adding random access and interactivity makes the material more useful and accessible than it had been in its previous medium.

Unleashing the Creative Impulse

One of the goals of good presentation is to make new things familiar and familiar things new. Often, this is a process of experimentation, trying different methods or different metaphors, until you strike the chord you are after. Obviously, the greater your access to dynamic content, the greater your options. Multimedia lets you choose the audio or visual that best suits your message.

In the beginning of the presentation process, you usually have time to look at various creative options and explore the best ways to communicate your message; but as time runs short, creativity becomes less and less welcome. In the nondigital past, this has been a problem. For example, if you are working with traditional linear videotape as your delivery medium, you endeavor to design creative scenes and camera angles. You might try several versions of the same scene, looking for just the right result. In the editing bay, experimentation continues with the sequences you assemble and the effects you add, until a final edit decision list is written. At that point, you are ready to make a master tape. The chance to get creative and do a bit of free thinking and rethinking has passed, or at the very least has become much more costly. If the worst happens and certain parts of the presentation bomb, the only option is a new edit and master. The same scenario applies on a lesser scale to creating a 35mm slideshow.

Now, with digital multimedia, the creativity window is open far longer and much wider. Experimentation becomes not only less expensive and easier, it becomes almost irresistible. No slides need to be made and no videotapes need to be recorded. Images created in the computer can be altered, changed, and exchanged, new music can be substituted, and even video can be quickly recut or replaced.

Multimedia Rationale

Only four reasons justify adopting any new system, technology, or business strategy: you must be able to do something faster, cheaper, easier, or better. Multimedia presentation techniques and technology offer all four, though not necessarily all at once.

Faster

Whether it was Benjamin Franklin or the inventor of the Rolex wristwatch who first said "Time is money," the phrase is nowhere more apt than when applied to presentation production. Time invested in preparing any type of presentation can run the gamut from hours to days to months, and it must be balanced against the projected return. While the benefits of a powerful, well-designed presentation would logically be greater than the benefits from a bland and simple presentation, the time necessary to develop the former is not always justifiable. The level of quality and impact is a direct function of how much time is available.

Multimedia technology and techniques allow the creator of a presentation to save time by introducing dynamic sound and motion more quickly than is possible in a nondigital environment.

With the proper tools and the right content, creating a presentation with sound and video clips can be as quick as creating a presentation that uses only black-and-white overhead transparencies for speaker support (though learning to use the multimedia tools may require more time initially). Adding motion and sound to a presentation once meant time-consuming recording and editing. Magnetic tape remains the most convenient and cost-effective way of capturing the initial sound or video, but nonlinear, digital editing and digital storage dramatically reduce the time spent in the editing room. (See Chapter 10.)

In general, content that is in digital form is far easier to alter. A digitally produced chart, for example, can be altered and updated with a few keystrokes and mouse clicks, whereas the same chart drawn and painted on artboard might have to be completely redone. New text, graphics, sound, animations, and video can often be added or substituted by simply changing a source file. Tinkering with the length of an animation, sound, or video clip can be done without destroying the timing of the rest of the presentation. Entire segments of a presentation can be swapped out, reedited, or deleted.

Cheaper

Multimedia, while more costly in the initial stages of system investment and learning time than some other presentation tools, can bring down overall costs in a number of ways. The most obvious savings are realized in the elimination of the costs associated with producing 35mm slides, videotapes, or other analog delivery materials, by delivering the presentation directly from the computer. The savings are multiplied by the number of sets of materials that are distributed to presenters throughout the company. The reduction in editing time, mentioned earlier, also translates into bottom-line savings. Once an image is created in digital form, it can be used over and over again, eliminating extra development costs.

By far the most significant dollar benefits come not from reduced production costs, but from enhanced communications capabilities. Studies on the use of visuals, sound, motion, and interactivity conducted by organizations such as the Department of Defense, the Wharton Center for Applied Research in Philadelphia, the University of Arizona (see Chapter 2), and others have demonstrated dramatic improvements in communication. The gains in retention rates, learning speed, attention levels, credibility, and overall impact translate directly and indirectly into gains in productivity and revenue.

The Graphix Zone, a multimedia training and production company based in Irvine, California, estimates the following: a $200 million company, annually producing 50 slide presentations (presented 1,300 times) using in-house services and two seven-minute videos using an outside service, would realize savings of approximately $200,000 per year in hard costs, excluding hardware investment, if it converted to digital multimedia presentations. Factoring in a revenue increase of 15 percent as a result of improvements in communications, corporate image, and customer service, and assuming the company operates on a 15 percent gross margin, the value of the switch to multimedia presentation calculates to a total of $4.7 million.

Easier

Presentations delivered directly from the computer can make your job as a presentation creator dramatically simpler in several ways. First, assuming your visuals are created on the computer, leaving them there for presentation avoids the need to output to some other medium. While computer-based presentations have their own limitations, once a system is running and tested, it is relatively easy to use. As noted earlier, multimedia also makes building in and delivering audio and video elements in a presentation far easier. Setting up, cueing, and operating separate analog playback equipment is not necessary, and synchronizing the various media elements within the program is much easier than running several playback devices simultaneously.

Finally, multimedia technology can make it easier to communicate your message to your audience. As will be noted in several of the following chapters, some stories are better told with dynamic media. A 30-second video clip showing the procedure for fixing a paper jam in a copy machine might replace ten minutes of difficult description and several diagrams. Capturing the voice of a chief engineer describing a challenging technical concept would be easier than taking notes, asking questions, and trying to write the procedure in your own words.

It would be misleading to suggest that creating and delivering a multimedia presentation is always simple. In fact, top-level multimedia productions can be extremely difficult to create. What is important to understand at this point is that once you have put together a working system, you can add multimedia components to an onscreen presentation, at a basic level, as easily as you would add a chart or text. (See Chapter 5.)

Better

Doing anything faster, cheaper, and easier allows you to do it better. There would be no point in considering or even learning about multimedia presentation technology and techniques if they did not make you better at what you do.

Among the concrete benefits attributable to multimedia presentations are the following:

- Higher close rates
- More new business
- Improved corporate and personal image
- Enhanced credibility
- Improved order response time
- Better customer relations

The combination of making concepts clearer, using high-quality images, adding sound and motion with relative ease, making changes faster, and reducing production costs translate to more efficient communication and a more productive organization.

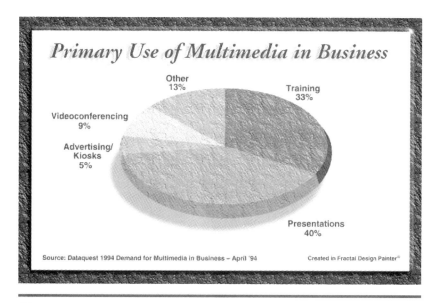

Figure 1-6 Primary use of multimedia in business

Multimedia Inroads

At this stage of development, the largest application for multimedia in business is presentations (see Figure 1-6). But you should not assume from this discussion that presentations are the only uses for multimedia. In fact, the technology has potential benefits in every aspect of business communications and information processing.

▶ *Training*, which is closely related to and often overlaps with presentations, is the second largest application for multimedia, and for good reason. The Institute for Defense Analysis determined that using interactive multimedia for training reduces costs by up to 64 percent and reduces time needed to reach equivalent levels of proficiency by 36 percent.

▶ *Videoconferencing*, already proven as a productivity tool, is quickly moving to the desktop, where it takes advantage of multimedia technology to let business people communicate visually, orally, and graphically from point to point. In a few companies, internal

Case Study:

Company: General Telephone Operations, Irving, Texas
Business: Voice, data and video products and services
Objective: Design and implement an interactive sales training and presentation system.

In 1993, GTE Telephone Operations (GTE Telops) determined that it had a critical and strategic need to provide GTE Telops sales personnel an effective means to present the facts and business advantages of the company's broad range of communications services. The company was rapidly producing new products, services, and pricing configurations and realized the need for faster solutions that allowed greater customer participation.

GTE Telops also realized how costly it is to pull sales staff out of the field for training and that the company could be missing significant sales opportunities in the process. GTE Telops needed a way to help its customers learn the facts and benefits of its new products while maximizing the resources of its sales force.

The solution, GTE Telops determined, was a portable, interactive, multimedia information system. The company already had foreseen the po-

tential of multimedia. In the late '80s, GTE's internal video broadcasting and production facilities were consolidated into GTE VisNet, which was chartered to explore viable business applications for video and interactive multimedia.

Although VisNet currently produces video and multimedia communications services for a wide variety of businesses and institutions, it also utilizes external resources when specific expertise is needed. For this project, VisNet contracted with Dallas-based BERMAC Communications. BERMAC worked closely with VisNet to develop options for an integrated information solution that would meet GTE Telops' objectives.

The result was the GTE SalesWare Project: an integrated multimedia sales presentation system that combines sales training with a dynamic presentation architecture to facilitate consistent, high-quality presentations to customers. The GTE SalesWare project initiated a number of the key steps that clearly illustrate both the process and the potential of multimedia presentation development.

The project began with a thorough information gathering effort by BERMAC and VisNet. Their challenge was to determine the specific

products and business issues to be covered, the information necessary to present these products, and supporting capabilities required by the sales force in both hands-on product training and practical presentation of the information.

The development team identified a need to present background and company context information, technical background and specific telephone service detail. In addition, the team determined that cost advantages of certain programs would be more effectively showcased by real-world, case study situations that could be integrated into the presentation.

It also became clear that presentation of the basic information was not sufficient to drive the sale. The program had to have components allowing the salespeople to tailor the information to specific customer needs and to interactively price GTE products and services in real time.

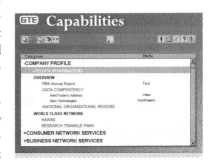

GTE Telops

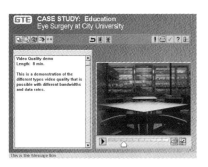

Finally, it was determined the system should function as a two-way information solution. This design would allow information to be captured from the prospect and interactively analyzed in the system to select the most relevant and useful services for that particular customer.

BERMAC and VisNet advised GTE Telops to select portable presentation computers based on the Intel 486DX processor and including all the necessary multimedia hardware and software to play back full-color images, sound and full-motion video.

The presentation was designed as an interactive software environment that could be operated just like a standard menu- and icon-driven software program. A series of operating icons were placed off a main screen that allows the sales executives to access any of the information components on demand, including product features, capabilities, availability, and pricing. A "quick quote" feature allows the salesperson to calculate instant price quotes based on customer data.

Using easy-to-read display lists of topics and information, the system lets the user select and navigate to the desired information detail, presented in the form of graphics, text, data tables, or full-motion video. Larger topics, such as the GTE company profile, are segmented into on-demand subheadings so that the presentation user can quickly learn facts and access supporting background materials, including video speeches by GTE executives.

The project can be operated by GTE Telops field contact representatives in large or small group presentations. In practice, the design has proven so successful and user-friendly that customers can query the system themselves. They can, for example, experiment with different service and configuration options at their own pace, without the salesperson present.

Overall, the GTE SalesWare system lets the company intimately and directly address specific customer needs in a way never done before. Ed Gillenwater, VisNet's manager, advanced communications, calls this "moving from a share-of-market-driven sales approach to a share-of-customer focus." Instead of pushing customers into broadly defined solutions, GTE SalesWare allows them to participate in developing "customized" product solutions.

From GTE Telops' standpoint, project manager James V. Coleman says, "The presentation system meets several practical objectives, including shortening the overall sales cycle and increasing credibility with business prospects."

Multimedia technology lets the system combine sales training and presentation material into a unified system, eliminating duplication of efforts to create the materials. While this does not lessen the need for well-trained salespeople, it ensures every salesperson access to the answers and facts they need.

Because it is not a stand-alone presentation, the system offers the opportunity to integrate its internal systems for storing and presenting market and product information with the systems used for customer presentation. According to GTE Telops, this project allows a consistent platform between internal and external communications, as well as providing a dynamic, multimedia framework for information management within the context of a fast-evolving and expanding marketplace.

e-mail has already converted to using multimedia; and as networks and the communications infrastructure improve, the same will be true of all electronic mail.

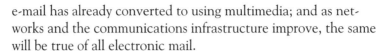 Multimedia *databases,* which will one day be as common as numerical and textual databases, allow information to be stored and accessed in whatever medium is most useful. Many real estate and travel businesses, for example, already employ multimedia databases (and presentations) to give salespeople and clients an audiovisual preview of properties and destinations.

Multimedia *documentation,* including manuals and reference books, put the information they contain into the most useful and beneficial form for the user. Manufacturers have begun to put interactive multimedia manuals on the factory floor to assist maintenance personnel. A few companies have been experimenting with putting company policies, standards, and procedures in kiosks using multimedia documentation instead of in hard-copy document form.

The very fact that you are reading this book makes you a member of an exclusive club. It demonstrates your interest in the communications benefits of multimedia and its evolution as a presentation tool. Not only are relatively few organizations now using any form of multimedia technology, the overall demand for multimedia capability is still relatively low. The respondents to a 1994 study conducted by Dataquest, a San Jose, California–based market research company, revealed that overall demand for multimedia capability has yet to take off. Most (51 percent) said the demand in their companies was low. Another 31 percent characterized interest as medium. Only 16 percent said the demand was high.

Within the minority of companies that are using some form of multimedia, widespread usage among employees is rare. Surveying companies with more than 1,000 employees, Dataquest found that in nearly half of the companies, ten or fewer individuals are using multimedia of any kind. Approximately three-quarters of the companies surveyed have 50 people or less working with multimedia. Only 1.5 percent have expanded multimedia usage to more than 500 people.

Nevertheless, beneath the surface the pressure is building. Within corporate graphics departments, spending for multimedia-related products is rising swiftly. Even in companies that have not yet committed to multimedia presentations, in-house graphics professionals, already aware of the potential and the trends, are equipping themselves for multimedia production.

Industry analysts estimate that within two years, nine out of ten personal computers sold in the U.S. will be multimedia-capable. While that still leaves a huge installed base of machines that must be replaced or upgraded for multimedia use, the technology trend is a clear reflection of what can be expected at the applications level.

If your organization has already committed to multimedia presentation or at least begun to explore its potential, you are at an advantage. But the advantages of early adoption will be remarkably short-lived. Trends observed in the evolution of other business communication productivity tools (the facsimile machine, for example) suggest that once the benefits of multimedia presentation become clear (as the result of the work of pioneering individuals and companies), the rest will rush to bring their capabilities up to date. In the long run, multimedia in general and multimedia presentations in particular will become a way of life for business.

Our Digital Future

The convergence of the computer, entertainment, publishing, telecommunications, and consumer electronics industries has crowded the headlines for more than a year and has become a footnote in the discussion of the future of almost any business. This convergence is rooted in the notion that *all* information will eventually be collected, stored, accessed, and transmitted in digital form.

The digital synthesis of information is foreseen as a soup-to-nuts solution, which includes all of the business and technology components necessary to create and develop content and business applications, distribution and deployment of the applications, and the necessary devices for users to access and benefit from them.

Businesses will be forced to adapt and adopt digital information technologies or perish. The value in information will be less in its ownership than in its timely access and analysis combined with equally timely distribution. In an all-digital business environment, then, the use and access of information reaches all the way from its source down to the personal computer on the desktop. Soon, no significant distinction will be made between accessing numerical data, text, graphics, sound, or video; the lines of communication will be open to real-time exchange of dynamic media (see Chapter 11).

The changes wrought by digital convergence affect the design of the computer and network system, the telecommunications system, office productivity devices, meeting style, and even business location. In the all-digital business world, the creator of a presentation looks for content wherever and in whatever medium it exists. It may be found in the company's databanks or it may be in the Library of Congress or anywhere out on the web of the National Information Infrastructure (NII), the so-called information superhighway being championed by the current Administration.

Because the content used to build a presentation will arrive at the desktop in digital form and will leave the same way, presentations in an all-digital environment need no longer be stand-alone events. When desired, the content will be linked to live data feeds, changing just before or during the presentation. Information generated during the multimedia presentation will enter the system and become part of the database, and as it does the presentation content in all other areas of the business will change accordingly. If so programmed, the new information may trigger other actions or events, such as scheduling a product shipment, updating inventory, or sending an advisory to the product design department.

To some extent and on a small scale, integrated information systems, including interconnected presentations and training systems, are already being implemented. For most companies, such high-concept information systems are still years away, but the time to start thinking about information in a new way is already here.

Freedom of the press belongs to the person who owns one.

—A. J. Liebling

MULTIMEDIA AND THE ART OF PERSUASION

Presentation is a way of life in every business; it's almost a ritual in some. Though the tools, techniques, and styles may vary, presentation is integral to all levels of commerce, from the bazaar to the boardroom. The integration of multimedia into the rites of presentation combines contemporary technology with the timeless presentation objective: persuasion.

Presentations and Corporate Culture

Presentations in business take many forms. They can be internal and external, formal or casual, one-on-one, one-to-many, or many-to-many. Presentations can be evangelistic, designed to promote a person or concept. Or, they can be pragmatic, designed to impart specific information in a persuasive fashion.

The types of presentations created and delivered in an organization generally depend on the size and nature of the business itself. In business, as with individuals, the process is largely one of personal style. Where some organizations work diligently to structure and refine their presentation capabilities, others let the process develop organically. While some companies maintain highly organized systems to govern the authoring and delivery of presentations, others adopt a laissez-faire approach that leaves individuals to their own devices, literally. There is something to be said for both approaches. The first serves to homogenize the message and present a unified, consistent company image. The second allows the talents and temperaments of individuals and departments the freedom to work to their strengths and proceed as they see fit.

The growing arsenal of presentation technology provides opportunities for companies with buttoned-down collars as well as those with rolled-up sleeves. Indeed, thanks to advances in personal computers, peripherals, and display technology, the options are so great that many companies find themselves struggling to finalize buying decisions and establish presentation guidelines.

Presentation Applications

Multimedia technology has dozens of current and potential applications. It is not exclusive to any one type of presentation. It can be applied to greater and lesser degrees, according to the needs of the information. In some instances, multimedia is nothing more than a bit of audio for emphasis at a desktop meeting. Other times, multimedia is the pounding rhythms and dazzling visuals on a 30-foot-high videowall during a new product rollout. Multimedia has also created new forms of presentation, such as interactive sales and marketing brochures.

It is difficult to find any business-related activity involving information and persuasion that is not a candidate for the impact and enhancement of multimedia. Possible events include

- Speeches and lectures
- Portable sales presentations
- Boardroom presentations
- Annual meetings
- Dealer and sales meetings
- Financial briefings
- Work group sessions
- Training sessions
- Product briefings and introductions
- Press conferences, conventions, and trade shows
- Information kiosks

Keep in mind, many of these applications overlap. Presentations created for one event are often of use for another. As pointed out in Chapter 1, one of the advantages to using multimedia is the ease with which multimedia content can be altered, repurposed, and customized. A complete presentation strategy looks at not only what multimedia can do for a particular application, such as sales presentations, but also at how production and distribution of multimedia materials can provide economy and efficiency by serving a variety of different information and communication functions.

Experiencing Information

Presentation in general and multimedia presentation in particular are not about information dumping. Hauling information from one place to another and shoveling it at an audience is a chore best left to words or figures on a sheet of paper, on a disk, or perhaps via E-mail. The primary mission of presentation is persuasion. To persuade, you must design your presentations so that your audience has an information experience.

Studies designed to find out how people learn and remember have estimated that we retain about 20 percent of what we see, 40 percent of what we see and hear, and 75 percent of what we see, hear, and do. We need nothing more than our own experience to know that we learn faster, comprehend better, and retain more when our senses are stimulated on multiple levels.

Multimedia's greatest asset as a presentation tool is its broad engagement of the senses. When applied to presentations, multimedia addresses primarily the senses of sight and sound, though in the case of interactive kiosks, it also includes some degree of touch (doing) as well. Multimedia capitalizes on human psychology and physiology. It reaches the mind *and* the heart of an audience via the senses. When used correctly, it produces more than an intellectual response to the information. It creates a positive information experience targeted specifically at promoting your message.

Consider sight. Seeing is obviously far more than capturing and focusing mental pictures of the physical world. Seeing is a form of sensory reasoning in which the brain receives a constant stream of millions of signals and must quickly translate them into a coherent picture of the physical world. Even the most impassive researchers continue to be astounded by the ability of the brain to construct or invent a visual understanding of the world almost instantly. We see color, detect motion, identify shapes, gauge distance and speed, and judge the size of objects, literally in the blink of an eye. In addition, the brain is able to make corrections and adjustments to flawed or incomplete visual information.

The human eye is by far the most powerful information conduit to the brain. The retina of the eye contains 150 million light-sensitive rod and cone cells and is actually an outgrowth of the brain itself. Neurons (nerve cells) devoted to visual processing number in the hundreds of millions and take up about 30 percent of the cortex of the brain. This compares to 8 percent for neurons devoted to touch and only 3 percent for hearing. Each of the two optic nerves, which carry signals from the retina to the brain, consists of a *million* fibers. Each auditory nerve carries a mere 30,000.

With all that bandwidth to the brain, no wonder the majority of information we take in comes through our eyes. We can read five times as fast as the average person can talk. We can register a full-color image, the equivalent of a megabyte of computer data, in a fraction of a second. When we watch a video or a movie, we are seeing about 24 to 30 megabytes of visual information per second. From that data, we are able to distinguish color, light, and movement as well as spatial depth. Not bad for a 3.5-pound, 12-billion cell processor.

Multimedia takes advantage of the available cerebral bandwidth by addressing more of it. Sights and sounds combine to reach more areas of the brain and trigger a deeper, more thoughtful response. In his book, *You Have to Be Believed to Be Heard*, communications consultant Bert Decker cites scientific evidence that suggests our deepest instinctual responses, perceptions, and emotions stem from the most primitive part of our brains, an area he calls the "first brain." And in that area, he suggests, "gut" feelings determine such perceptual issues as believability, likeability, and degree of importance. Which messages are accepted and which are rejected, then, depends more on subliminal perceptions of sensory information than on conscious reasoning.

Whatever the physiological roots, the importance of creating a multisensory information experience is obvious to anyone who has compared the difference between the impact of a film strip and a 70mm movie with Dolby Surroundsound. You may not be able to fit the 70mm screen and projector into your conference room, but multimedia tools and techniques can carry your message much deeper into the minds of your audience than, say, a slideshow or a flipchart.

The Communications Imperative

The basic human need to communicate is at the heart of the presentation process and the information experience. Our personal lives, business lives, and social lives function on the exchange of information. We communicate to understand, to organize, to create, and to survive. While it may seem excessive to link business presentation with our very survival, the importance of persuading and motivating others cannot be overstated. Unless you are effective at reaching the emotions and feelings of others, in business and elsewhere, you will not be heard and your message will not get through.

Consultant Decker observes that "People buy on emotion and justify on fact." In the process of communication, if you reach another's intellect, you have a chance that your message will be "heard." If you reach their emotions, you have won.

The audience's emotional "buy in" that Decker and other presentation specialists speak of is based on the believability and credibility of the person and/or the organization presenting the information. For persuasive communication to take place, others must perceive that what you are saying is valid and important.

Persuasion and Perception

Perception, not reality, is the key to presentation. How an audience perceives the presenter and the presenter's message directly affects the importance they place on the information and what, if any, actions they will be persuaded to take. Perceptions are formed from a combination of the verbal (words), vocal (voice), and visual cues produced by the presenter. On both a conscious and a subconscious level, people process the cues through the filter of their own experience, and at that point perception becomes reality for the audience.

Contemporary businesses live and die based on how they are perceived. Companies spend vast amounts of time and resources grooming the perceptions of such groups as financial analysts, customers, the press, and employees. On an individual level, it is safe to say that almost any action you take in business will affect others' perceptions of you.

Of course, the closer perception and facts coincide, the more effective the message, but one is not necessarily dependent on the other. Take for example a company that has been repeatedly cited for violating environmental regulations. Through persuasive presentation by the CEO on the subject of a single, high-profile cleanup operation, the public and investors come to perceive the company as environmentally conscientious. Or, a company might happen to quietly spend millions of dollars supporting environmental causes, but an isolated incident of pollution, blown out of proportion by poor presentation of the issue, causes the company to be perceived as an egregious violator.

The nearly overwhelming challenge of managing perception stems in part from the improvement in communications technology that allows information to travel faster and farther. Information about an organization no longer flows through a single conduit and out to select receivers, but emanates (sometimes leaks) from dozens or hundreds of sources. In the final analysis, perception is managed by controlling which information is available and how it is presented.

Persuasion Process

Presentations occur millions of times each day, all over the planet, from the largest business theater extravaganzas to the briefest one-on-one sales calls. In each case, someone is trying to persuade someone else about something. Part art and part science, the process of persuasion involves every aspect of personal communication. From the personality and appearance of the presenter to the strength of the coffee served at the break, every component of the preparation, approach, content, and delivery affects how well or how poorly a presentation influences the audience.

To break down the process, start with the desired action or result and work backward, noting as you go all the potential variables in the persuasion process. Notice in Figure 2-1 that the action you desire your audience to take is based on the four primary components of persuasion: attention, comprehension, agreement, and retention. To some extent, success in each is required to ensure a successful outcome to the overall objective(s) of the presentation.

As the persuasion process model (Figure 2-1) demonstrates, multimedia presentation content will be most effective at enhancing attention, comprehension, agreement, and retention when it is designed with audience characteristics and various fixed factors in mind. Chapter 4 discusses at length the effect of audience, content, and presentation environment variables on multimedia presentation design and planning.

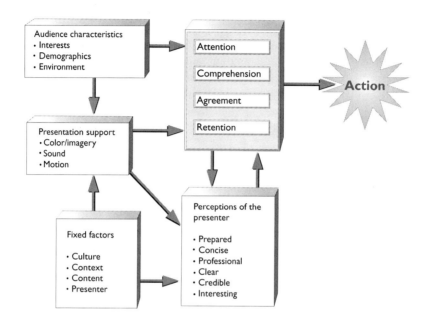

Figure 2-1   Persuasion Process Model

Attention

To persuade someone to take action, you must first gain and hold their attention. In the days of the MTV generation, the sound bite, and the video clip, attention has become a valuable and sometimes scarce commodity. Not surprisingly, the average attention span of a child has been estimated to be only about six seconds; children, after all, do not focus on any one thing for very long and are highly susceptible to distraction. What you may find surprising is that the average adult's attention span is just two seconds longer. Basically, that means if the members of your audience are not given fresh stimulation with interesting ideas or engaging stimuli, you will lose them in about eight seconds. You may be able to bring their attention back at some point, but every time they begin to doodle or daydream, you are losing impact and valuable persuasion time.

Comprehension

Knowing if and when someone comprehends what you are telling them is difficult. Some of the most glaring errors in human communication result when one party assumes that the other has comprehended and proceeds to act on that assumption. Ralph Waldo Emerson said it best when he observed, "It is a luxury to be understood." Nowhere is that more true than when you are attempting to persuade. Depending on the complexity of the issue, comprehension may be total or partial. Levels of comprehension vary from individual to individual, and you should never expect that everybody is going to "get it" to the same degree. But you can be certain that if an audience is unclear on the concept, they will be unwilling to act.

Agreement

Bringing an audience into agreement with your point of view can be nearly effortless or devilishly difficult, depending on the predisposition of your audience and your powers of persuasion. An audience that wants to agree with your message will naturally give you every opportunity to make your point. They will perceive your arguments as favorable proof of their own conscious or subconscious desire to agree. A disagreeable audience, on the other hand, will be skeptical not only of your information and approach, but of you and your organization as well. Bringing someone into agreement can also be thought of as causing them to yield to your point of view through clear convincing argument. Keep in mind the advice of Dale Carnegie, "The best argument is that which seems merely an explanation."

Retention

Retention, the ability to remember information for some measurable period, is one of the most obvious components of persuasion. To be persuaded, an audience must retain at least enough of the message to facilitate their decision to act. When retention is combined with comprehension, the result is learning. Often an audience is in attendance to learn and the presenter fulfills the role of teacher; the presentation becomes the curriculum.

Persuasion and Technology

Multimedia provides a means to supplement a presenter's efforts to garner attention, increase retention, improve comprehension, and bring an audience into agreement. It can be used to further all four components of the persuasion process, operating both on a conscious and subconscious level. While multimedia is useful for such obvious purposes as explaining complex processes with animation or video, it is also particularly apt at reaching below the surface to effect persuasion on an emotional level, to provide a sensorial information experience.

In 1986, the same year the first dedicated presentation graphics software product was introduced, the University of Minnesota and the 3M Corporation conducted a study of the use of visuals in presentations. The study found that presentations incorporating visuals such as 35mm slides, overhead transparencies, and color graphics were found to be 43 percent more effective at motivating an audience to action than presentations that used no visual aids. The study also concluded that visuals increased retention rates by 10.1 percent, improved audience perceptions of the speaker by 11 percent, increased comprehension by 8.5 percent, boosted attention by 7.5 percent, and increased agreement by 5.5 percent (see Figure 2-2).

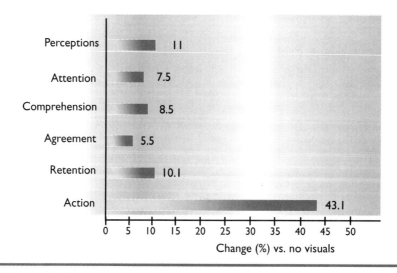

Figure 2-2   Impact of presentation support visuals (source: University of Minnesota/3M Corporation)

A later study by the Management Information Systems Department of the University of Arizona extended the investigation of the University of Minnesota/3M findings to the specific effects of dynamic, computer-generated presentation visuals, such as transitions and animations. That research compared the impact of static visuals (slides and overheads) and dynamic visuals (animations and transitions) against using no visual support at all. Among its most telling conclusions: overall perception of the presenter improved by an average of 16 percent when animations and transitions were used in an onscreen presentation, but only by 6 percent when static visuals were used to convey the same message (see Figure 2-3).

Compared to presentations using no visual support, presenters were judged to be nearly 40 percent more professional, 11 percent better prepared, 10 percent clearer, 5 percent more concise, and 24 percent more persuasive. Using static visuals for speaker support, presenters were judged only 10 percent more professional, 18 percent more persuasive, and 2 percent clearer. In the study, the audience perception of the

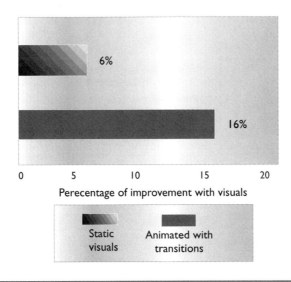

Figure 2-3   Overall perception of presenter (percent improvement) (source: University of Arizona–3M Meeting Management Institute)

presenter actually slipped in preparedness and conciseness ratings when using static visuals (see Figure 2-4).

The University of Arizona study also concluded that multimedia enhancement increased an audience's agreement with presenters by 17 percent, boosted credibility by 27 percent, and improved attention paid to the presenter.

One particularly revealing aspect of the study compared presentations by two different presenters, one of average presenting ability, the other above average. When the average-ability presenter used multimedia technology, ratings for perception, comprehension, and agreement were equal or better to the ratings achieved by the above-average presenter when the latter used no visual support materials. For above-average presenters, the study found that not only does multimedia improve their persuasive impact, but the audience *expects* presenters of their caliber to use high-quality visual support, and their credibility ratings suffer without it.

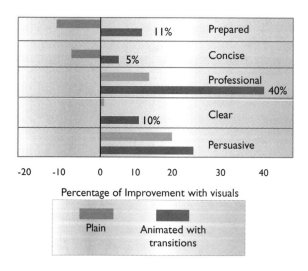

Figure 2-4 Perception attributes (source: University of Arizona/3M Meeting Management Institute)

The research only confirms what most presenters understand instinctively: you will be more persuasive when using well-designed, high-quality, internally consistent visuals to support your message.

Presentation Tools: A Brief History

As noted at the beginning of Chapter 1, the process of collecting, organizing, and presenting information for the purpose of persuasion is as old as human communication itself. But technology designed specifically for enhancing the persuasion process played a relatively small role for the first 20 millennia or so. Some of the earliest tools, such as the pointer stick, chalk and slate, and whiteboard and pen, have endured to this day. The invention of movable type by Gutenberg in the 15th century may have been a boon to speaker handouts, and engraving may have been a useful way to mass-produce early presentation graphics, but other than that, presentation technology changed little until the last 100 years or so, and enormously in just the last ten years.

Evolutionary Fast Track

The pace of scientific advance began to pick up rapidly in the last century, and with it new options appeared for presentation support. Not long after the first experiments were conducted in silver halide film photography, the medium was adopted by business as a means of showing off products that were too large or too heavy to transport. Presenters were among the first to use candle-lit slide projectors, and later they were among the earliest to embrace the motion picture camera for its ability to communicate a sense of reality and presence.

Still, progress was slow until very recently (see Figure 2-5). The ubiquitous overhead projector and carousel slide projector, for example, have been the staples of electronic presentation technology for more than 30 years and continue to be the most commonly used electronic presentation display devices (not counting laser printers for handouts). With the advent of magnetic audio tape, presenters were able to more easily record and edit sound for use in presentations. In the 1970s, videotape made full motion a cost-effective option. In the early 1980s, videodiscs were adopted as presentation tools (see Chapter 10).

Case Study:

Company: Union Bank of Switzerland, North American Region, New York

Business: Provider of commercial, merchant, private, and investment banking services

Objective: To produce a flexible, in-depth presentation for senior management, demonstrating the power of multimedia technology as a communications tool

At the North American regional office of Union Bank of Switzerland, internal boardroom presentations to senior levels of management are typically created using basic presentation software. Static lists, charts, graphs, and text are output to either 35mm slides or overhead transparencies for delivery. Vice president of marketing and corporate communications, Nathan Stack, had for some time been investigating the potential benefits of multimedia for internal and external presentations. But he realized he faced a difficult challenge trying to come up with a way to convince a large, conservative organization that multimedia technology should be added to its presentation assets.

Stack spotted his opportunity when he was asked to make a presentation to his counterparts from the bank's worldwide divisions at corporate headquarters in Zurich. He saw it, he says, as an opportunity to make the medium his message.

Each of the division executives had been asked to prepare a presentation outlining their marketing and communications strategies. Stack chose to share his division's exploration into the application of digital media. "What better way to show

them the concept of digital media than to put it up on the wall," he says.

Aware that his resources and experience with multimedia were limited, Stack went outside for help. He hired the Burnett Group, a Manhattan communications agency, to design a multimedia presentation that would allow him to showcase the new technology. The Burnett Group created the presentation on a Quadra 800 using Macromedia Director, Specular Infini-D, Adobe Photoshop, and Adobe Illustrator.

Union Bank

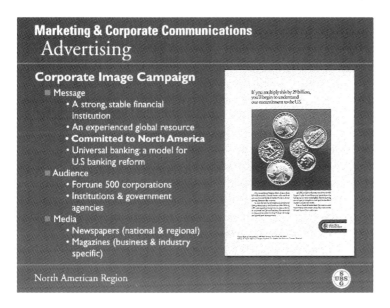

Marketing & Corporate Communications
Advertising

Corporate Image Campaign
- Message
  - A strong, stable financial institution
  - An experienced global resource
  - **Committed to North America**
  - Universal banking: a model for U.S banking reform
- Audience
  - Fortune 500 corporations
  - Institutions & government agencies
- Media
  - Newspapers (national & regional)
  - Magazines (business & industry specific)

North American Region

Stack arrived in Zurich with a Macintosh PowerBook 180C and a Proxima Ovation LCD overhead projection panel. He opened with an animated logo that morphed into a pie chart and served as the presentation's main menu. By clicking on a particular piece of the pie, Stack could launch interactively into any portion of the presentation where various animation and transition techniques were designed to enhance the information and dazzle the executives. Each section was several layers deep, allowing him the flexibility to respond to the varying interests of his audience. Stack admits he did not know exactly what to expect from his European counterparts. How much did they know already? What were they most interested in? "It was a great comfort knowing I had a number of different materials available to support my point of view," he says.

Because Stack knew he would be the only presenter using the program, the Burnett Group avoided cluttering the screen with buttons and arrows. They hid the interactive controls from view until the cursor would pass over them, at which point they would appear to tell Stack he was clicking on the right control.

Though the presentation was modest by some multimedia standards, it served its intended purpose. Says Stack, "Most of the questions I got afterwards were about how they could start doing these types of presentations in their regions."

Presentations combining a variety of analog media types (audiotape, videotape, 35mm slides) became known as *multimedia events,* and they gave the business communicator previously unavailable powers to persuade and influence. Multimedia shows soon became the presentation top guns, proving their ability to target an audience and strike with persuasive force. But the complexity of synchronizing multiple types of audiovisual equipment and the need for expensive creation and delivery talent relegated multimedia to special event status. The multimedia

MULTIMEDIA — 1990

Digital video	1989
High-res computer grahics	1988
Desktop publishing	1985
Databases (text-based)	1982
Spreadsheets word processing	1980
Videodisc	1974
Kodak Carousel slide projector	1961
Videotape	1956
Overhead projector	1944
Television Moving pictures	1926
Film projector	1887
Fountain pen	1880
Telephone	1876
Photography	1822
Chalkboard	1700s
Movable type	1476
Paper	105 A.D.
Sumerian alphabet	4,000 B.C.
Cave painting	17,000 B.C.
Stick Pointer	Pre History

Figure 2-5   A technology timeline shows the glacial evolutionary pace of communication (and presentation) tools until a sudden acceleration corresponding to the development of the personal computer

experience was limited to a relative few. Even today, shows incorporating multiple forms of nondigital media are still reserved for high-profile, big-budget special events.

The Debut of the Personal Computer

The personal computer began its career in the late 1970s processing numerical data. Like a teenager munching popcorn in a movie theater, the personal computer would happily crunch numbers as fast as they could be fed in. When the time came to change the personal computer to a word processor, converting words to letters and letters to numbers was relatively simple, so the computer could keep on doing what it did best.

By the mid-1980s, personal computers were being used to start and stop the playback of VCRs as well as control the operation of slide projectors and audiotape players. Coupling a videodisc player and a computer, a presenter could, for the first time, randomly and rapidly access video and audio segments. The freedom to control the flow of information was intoxicating. Typically, however, computer control required the expertise of experienced programmers, often using low-level programming languages to scratch out the software code line by line. The new-found freedom was not really freedom at all.

While all this was going on, the personal computer was busy revolutionizing print communication with powerful new desktop publishing hardware and software. Then, in 1986, with the introduction of a software program called More 1.0, presenters began to have access to the power of the personal computer for the creation and display of text, charts, and graphics. Though it was a black-and-white graphics program and not very easy to use, More 1.0 was a breakthrough for presentation because it could generate a series of bullet charts from an outline. The images created in More 1.0 and the other early programs were (and still are) intended as output to 35mm slide film and overhead transparencies rather than for computer screenshows. Nevertheless, presentations had gone digital.

The first presentation software programs were awkward and complex. The learning curve was not only steep, it took the user through a few loop-the-loops as well. But the improvements began arriving every few quarters, until the packages had earned themselves a place on the office shelf next to the spreadsheet, database, and word processing programs. From there, the pace of development accelerated almost beyond the ability of the business community to keep up.

Tools for digitizing sound and video gave presentation software access to dynamic media types, leading rapidly to full-fledged multimedia presentation authoring packages (see Chapter 5). Dozens of toolmakers joined the rush to bring multimedia presentations into the mainstream. The end result is a spectrum of products, both hardware and software, for producing and delivering presentations (see Chapter 3). Of even more importance, the products target a wide range of users, from amateurs to professionals.

Where once presentations required the work of dedicated departments or outside services, multimedia can now be produced at the desktop. The evolution in technology has transformed and continues to alter the way presentations are produced within the corporation. If you have not already felt the impact of digital communications technology, multimedia presentations included, you will soon.

Multimedia Presentation Strategy

Logically, those organizations that have made strong commitments to presentation resources in the past have been the earliest to adopt multimedia, but there are as yet few, if any, widely accepted models for incorporating multimedia technology into the business. Some early adopting companies have emphasized the use of multimedia in training, but they have given little attention to the technology as a presentation tool. Others have gone straight to full-scale portable sales solutions. Still others have reserved the impact (and expense) of multimedia for special events.

As noted in Chapter 1, the early adopters within an organization are often a small cadre, perhaps a lone individual, setting out to explore multimedia without the guidance of a corporate strategic plan. The rapidly changing products and shifting prices have made it difficult for many organizations to discover what is available, much less how best to implement it. Determining your approach as an individual or as an organization involves identifying what types of presentation you already do and what other types you would like to do or plan to do.

As you learn more about multimedia presentation, including the tools and their capabilities, you will need to develop a plan for adoption and implementation. There is nothing unique about strategizing multimedia. The principles are the same for the adoption of any new production or management system. The steps apply equally to using multimedia for individual productivity or embracing it as a company-wide communications strategy. First you evaluate your needs, then you investigate the potential solutions, then you initiate.

Evaluate

The fact that you have invested in this book probably indicates that you have evaluated your communications needs and found areas where you think performance could be improved. Perhaps you are dissatisfied with the close rate on sales calls, or the lead generation at trade shows is weak, or you have been hearing snoring in the back of the room during your presentations. Take note not only of recognized deficiencies, but also of new opportunities to connect with your audience. As metioned earlier, interactivity introduces a dimension to presentations that few companies or individuals have yet to explore.

Investigate

This book may be your first step at investigating the multimedia solutions available now and on the horizon. If so, use it to give you the necessary overview of the technology and what it can do before you delve into specific solutions. Because multimedia can involve complicated technology, production, and design, many vendors and service providers are offering turnkey (hardware, software, and production) multimedia solutions. While this may be the best way to start, you will not be as adept at making an informed decision unless you take some time to understand multimedia presentation technology in context. You should understand, for example, the repercussions of committing to a particular computer platform (see Chapter 3), or what the various levels of digital video quality mean in terms of hardware compatibility and cost (see Chapter 10).

Initiate

Once you reach a decision to adopt multimedia, have a plan. Perhaps you will decide to start small, adding sound or animation to a screenshow using presentation-capable software that is already on your shelf. Or, perhaps you will decide to invest in one of the newer programs created specifically to author multimedia (Chapter 5). You might start by con-

Presenter Support

Successful implementation of multimedia requires commitment and follow-through to ensure that it is used to its best advantage. All the glorious promises of better communication and greater impact will be lost if the presenters/users are not trained and supported. Later in this book (the Case Study in Chapter 10), you will read about the state-of-the-art multimedia sales presentation system implemented by First Data Corporation. In addition to being an outstanding example of implementation, the First Data program demonstrates the importance of presenter support.

Recognizing the need to maximize the potential of its multimedia sales presentation and avoid embarrassing system failures, the marketing communications staff at First Data prepared a manual to accompany the 11-part presentation, including a word-for-word script. The documentation includes instructions for the operation of the notebook computer as well as procedures for connecting it to a video or computer monitor.

The presenter is also taught how to hook up the external speakers and the hand-held mouse. In addition, the company has set up an arrangement with a rental company to provide any additional equipment that may be necessary while on the road, such as a presentation monitor or projection system. Salespeople are provided with the name of a specific person, who has been briefed on exactly the type of equipment required for the presentation. As a means of follow-up, the presenter materials include a survey to be given to the client at the conclusion of the presentation. While the presentation was created by an outside service, First Data has also hired an in-house multimedia designer to design some of the screens, oversee the production, and guide the company in the use of multimedia for future projects.

tracting out a complete presentation, with the idea that you will learn the process and bring it in-house at some point (see Chapter 12).

Human Factors

Multimedia presentation involves technology *and* people. As with any technology, some individuals take better to using the new set of tools and multiple media types than others. Rushing ahead with tools too advanced for the presenter is far worse than being slow to implement multimedia. If you are in search of presentation products to enhance your own persuasion skills, be honest with yourself about your computer skills as well. Many of the products mentioned in Chapter 3 and elsewhere require at least a basic working knowledge of computer hardware and software installation. You will need to be familiar with such basic computer operations as storage and file management. When dealing with display and other external devices, you will run across issues of cabling and compatibility that require time and effort to understand.

If you are evaluating the use of multimedia presentation for others, you should take a careful look at the skill levels and capabilities of the individuals who will be making the presentation. A person who is uncomfortable coordinating the advance of 35mm slides, for example, will likely have an even worse time with a mouse and onscreen interactive buttons. Watch out for hidden fears or phobias. It is a well-documented fact that many people are apprehensive about using new technology.

Technophobia

In his book *Sein Language,* comedian Jerry Seinfeld notes that most surveys have found that people's number-one fear is public speaking. The second-greatest fear is dying. So, according to those results, says Seinfeld, at a funeral you are better off being in the casket than giving the eulogy.

The impact of technological evolution on people has always been a thorny issue. Some individuals would not accept the automobile as a replacement for the horse. Some individuals to this day will not fly in an airplane. Occasionally, you will make a phone call and will receive no answer (a likely case of techno-rebellion against the answering machine).

Granted, if man was meant to fly, then the airline food would probably be better. But even so, the reasons for rejecting a new technology are for the most part irrational, born of a fear of the unknown or the unfamiliar. Only when one is comfortable with the tools can one make a presentation with confidence.

The behavioral components of presentation technique have been studied and discussed in dozens of books, magazines, and videos. Each comes to roughly the same conclusion: the presenter's skills and attributes are the most important aspect of any presentation. The checklist given to every presenter includes such factors as appearance, gestures, posture, vocal inflection, confidence level, eye contact, and so on.

The addition of technology, even while making the persuasion process more effective, places extra demands on the presenter. Here are two simple problems associated with a traditional 35mm slide show:

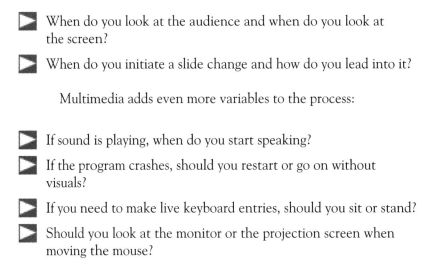 When do you look at the audience and when do you look at the screen?

▶ When do you initiate a slide change and how do you lead into it?

Multimedia adds even more variables to the process:

▶ If sound is playing, when do you start speaking?

▶ If the program crashes, should you restart or go on without visuals?

▶ If you need to make live keyboard entries, should you sit or stand?

▶ Should you look at the monitor or the projection screen when moving the mouse?

The key to becoming comfortable with technology is in finding simple and nonthreatening transitions from the familiar to the new, taking small steps and gradually building up to more ambitious projects.

Successful Techno-Human Interface

The following six behavioral guidelines address the most critical aspects of any well-delivered presentation. While they apply with or without multimedia support, they are the elements that are most often affected by the addition of high-tech techniques.

▶ *Gestures:* Nervous gestures and awkward hand movements can ruin an otherwise competent presentation. Use your natural gestures, not choreographed ones. Get used to the position of the mouse and keyboard, but do not hang on to them.

▶ *Smiling and facial expressions:* A sincere smile shows openness and enthusiasm. It says you are comfortable with yourself, your message, *and* your technology. Multimedia is exciting. Broadcast that message to your audience. If the computer or display bombs, do not let your smile bomb with it.

▶ *Eye contact:* Do not tell the audience that you are worried about your equipment by looking at it instead of them. Do not read text from the screen. If your images provide a visual cue, then you should pause, look to the visual, and then look at your audience and continue your presentation. Pausing is a very powerful technique. It shows you are confident and in control of the presentation. If you have to look down and use your keyboard or mouse, you should first pause and then continue by looking at the audience.

▶ *Posture and movement:* One of the first things people notice about you when you walk into a room is your posture and body movement. Standing tall and moving about lets you present yourself with energy and authority. You do not need to be anchored to a podium, and you should never hunker down behind the computer monitor. One of the nice things about using multimedia is the availability of a variety of wireless remote control options. With a remote control, you have the freedom to move about the room and interact with your media and the audience.

▶ *Voice and vocal quality:* Project your voice beyond the last row of the audience, and if you are using a lavaliere microphone, make sure it is balanced with the sound output of any media you are using. Never compete with your own support materials. Let sound clips finish before you speak. Do not speak over video that has sound. Do not describe what the audience will see unless it needs explanation.

▶ *Words and nonwords:* Ums, ers, and uhs commonly escape the lips of nervous presenters in the spaces between media elements, or when an unexpected lapse occurs. They are a bad habit and betray your nervousness to the audience. Pauses are perfectly acceptable. You should feel free to leave a two-second pause between thoughts in a sentence or between visuals without the need to make a vocal noise of some sort.

Testing, Testing, 1, 2, 3...

As a final note to achieving human and technological harmony, keep in mind that practice, always a crucial element of presentation, is doubly so when using multimedia. Become comfortable with the transitions, sounds, and video sequences in your presentation. It is the only way to be certain that all will go well and, therefore, the only way to present a confident attitude to the audience.

▷ Do not over rehearse; no less than three times but no more than six times is recommended.

▷ Do not make last-minute changes without a complete backup copy available.

▷ Do not lose your cool. As the saying goes, "Never let them see you sweat."

True Story

During the recent high-profile, televised California Economic Summit, ten minutes were budgeted to test out the computers and projection system for the electronic presentations. No rehearsal time was scheduled for the presenters at all. On the day of the presentations, a nervous but determined presenter insisted on controlling the advancing of his presentation from the podium rather than having an audiovisual assistant run the equipment from the back. A remote control device was quickly supplied.

The camera lights came on, everyone was watching, but the remote's transmitter was somehow blocked by all the electronic equipment in the room, and the presentation would not advance. The presenter became exasperated and, while the TV cameras rolled and the Governor of California looked on from the dais, the presenter let slip a familiar profanity followed by, "Can't, anybody get this technology to work around here?"

You get the picture.

The person who uses yesterday's tools in today's work won't be in business tomorrow.

—Anonymous

THE MULTIMEDIA TOOLBOX: AN OVERVIEW

A multimedia system is not one product, but a collection of sometimes dozens of interrelated components that must be carefully selected and configured to work together. The hardware and software tools you will need for your particular system depend on your budget, your skill level, and the complexity of the blueprints for the multimedia presentations you are planning to build.

Digital Power Tools

Unfortunately, at this stage of development, the new generation of digitally driven tools you will need to create and deliver a multimedia presentation do not come in one box. Rather, various products, both hardware and software, must be mixed and matched, plugged in and adjusted. Some products are as intuitive to use as a hammer or screwdriver, others are complex electronic devices with instruction manuals thicker than this book. Both general-purpose tools, useful for a variety of projects, and task-specific tools, designed for precise functions, are available. Understand from the start, you will not find a perfect solution or an ideal configuration. Every system will have its cost/performance tradeoffs.

The key to determining what you need to properly equip your toolbox lies in understanding how the components interrelate and what specific products suit the job you have in mind.

The products and their features change rapidly. Many of the software programs and hardware devices mentioned here will have been updated or replaced by the time you read this. Nevertheless, you should not let the array of tools or the lightning-fast evolution of the technology hold you back. If the products that exist today can make your business communications more productive, and if the return you expect is greater than the investment, then now is the time to begin assembling your toolbox. When you are ready to build a house, does it make sense to wait and see if someone invents a better hammer?

System Layers

Visualize the multimedia toolset as a series of concentric circles (see Figure 3-1). At the center are the central processor unit (CPU) and random access memory (RAM)—the brains of the outfit. From there, the circles radiate outward to include the operating system software, storage hardware, video display board, and the monitor. At that point, you have a working "computer" that only needs application software to make it functional. At the next level, you add the hardware for handling text, graphics, sound, animation, and video. Next, you introduce the media creation and editing software tools. Then come the authoring packages or media integration tools that put it all together. Next are the components for transporting and delivering your presentation. And finally, at the last ring, the presentation display monitors and projectors.

The layers or tiers of components are arranged, for the most part, in order of dependency. That is, the function of products in one circle depend on one or more of the products in the layers within. The onion-like structure does not necessarily tell you in what order you should buy products, nor does it suggest that you should start at the center and work outward. Building a system involves integrating tools that work well in concert. Just as a small hammer will not drive a large nail, a powerful authoring software package will not run on a slow or under-powered CPU. Likewise, using a costly and powerful personal computer to create and

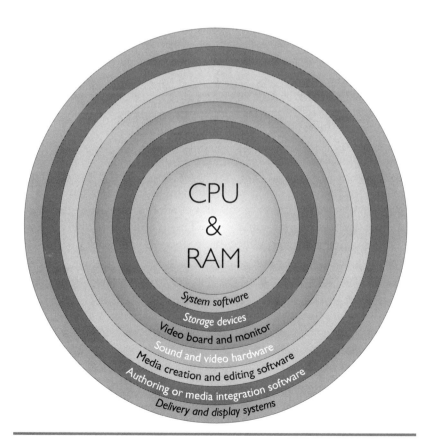

Figure 3-1   Multimedia system layers

deliver simple graphics and sound clips would be like driving thumbtacks with a mallet.

As you become familiar with the specifications and requirements of the hardware and software, you will begin to see how the performance of each tool, and the final quality of the output, is dependent on all of the other products in the system. For that reason, before you commit to any one product, it is best to familiarize yourself with all the categories. In addition, you should know to whatever degree possible what type of presentation you intend to produce. A complete multimedia system includes everything within the outside layer. But that does not mean you must have everything to begin. As will be stressed many times throughout this book, to find the best multimedia system solution, start by determining the needs of your business application and work backward from there.

This chapter will look at each successive layer of components, beginning with the core of the multimedia onion.

Development and Delivery

Multimedia presentation involves essentially two steps: developing and delivering. A *development system* includes everything you need to build a presentation from scratch. The computer (CPU, RAM, operating software, storage devices, graphics board, and monitor) is the foundation of a development system. To it you add the software applications and hardware tools for creating, editing, and integrating the audio and visuals that are your content (see Figure 3-2). The system on which a presentation is authored is also frequently the system on which the presentation will be delivered, but that is not always the case.

A *delivery system* encompasses the computer and all the required hardware for playing back your multimedia presentation, but it does not typically require the media building tools or even the authoring software (see Figure 3-3). A multimedia presentation can be delivered on a separate computer with hardware and software different from the development system, if the computer is powerful enough, the operating system is compatible, and the necessary playback components are installed. Each of the following chapters in this book will help you understand more about the distinctions between development and delivery systems.

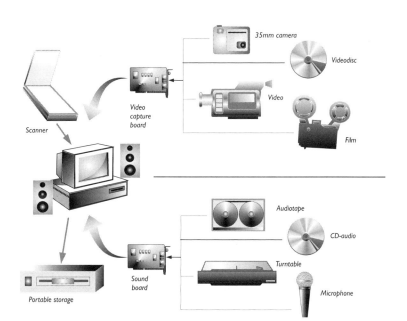

Figure 3-2 Multimedia development system

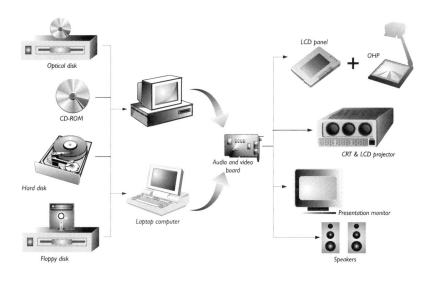

Figure 3-3 Multimedia delivery system

Core Layer: The CPU and RAM

The rigors of high-resolution graphics, video, and sound place extraordinary demands on computer processors. They add a tidal wave of data to the bit stream and increase the demand for calculation volume and speed. This is particularly true for the development system. Because this system is used to prepare and integrate the actual content for the presentations, it typically needs all the power your budget can muster.

Microprocessors form the foundation of any computer architecture. The power of a microprocessor to perform calculations depends on the number of transistors it contains and the speed at which it can perform calculations or operations. You will typically see CPUs identified by their clock speed as measured in millions of cycles per second or *megahertz* (MHz). The clock speed is the control pulse that synchronizes the microprocessor operations. The higher the clock speed, the faster the CPU processes data.

The Intel 486 chips, which contain 1.2 million transistors on a chip no bigger than a book of matches, are available with clock speeds of 25MHz up to 66MHz. Newer Pentium chips from Intel have a typical speed of 100MHz. Macintosh processors from Motorola include the 68030 and 68040 chips, which have clock speeds of 33MHz and 40MHz, respectively. The new PowerPC chip, developed by IBM, Apple, and Motorola, is not yet widely used for multimedia production; but its speed and power rank with the Pentium or better.

Though many of the software packages on the market will run on older 386 machines, if you are serious about multimedia production, you should consider as a minimum a machine powered by a 486SX chip running at 25MHz or better. On the Macintosh, a 68030 processor will do for basic presentations with graphics and sound, but the 68040 is preferable and will probably pay for the difference in saved time and frustration.

Random access memory, or RAM, is the place where your data is stored while the CPU is working with it. RAM is made up of banks of dynamic memory chips (DRAM), which are referred to as *volatile* because they hold information only while the computer is running; when the power goes off, everything in RAM is wiped out. In essence, RAM is where an application, such as an authoring program, is loaded and resides while you are authoring. It is also where the text, graphics, audio, animation, and video are stored as they are being created or edited. When RAM fills up, the data is sent to the hard disk for temporary storage, but the more the computer is forced to access the hard disk, the slower the performance.

The computer business has a saying: You can't be too rich, too thin, or have too much RAM. That adage is particularly apt for multimedia development, where the programs are large and the memory requirements for the media files often run into multiple megabytes. The 4MB of RAM included with most PCs will be enough to run some programs and create small media files, but will fall short when working with large graphics files, animation, video, or extensive audio. Faced with the large memory requirements of multimedia, you are advised to consider 8MB as minimum (professional presentation developers often equip their systems with 16MB or more).

The RAM demands on the delivery (or playback) system are ordinarily not as high as those for the development system. The computer is not required to handle multiple copies of the images and sounds, nor does it have to maintain one or more application programs in memory. A portable or desktop computer with 4MB of RAM may be all you need to play back a presentation that required 8MB or more to create.

Layer Two: System Software

The systems used to create and deliver multimedia presentations can be based on any of several platforms. Platforms refer to the type of processor that is built into the system. To make the processor and other components run, the computer also needs operating system software.

Platforms

The two most commonly used platforms for multimedia business presentations are the Apple Macintosh and what are frequently referred to as IBM PCs and compatibles (or *clones*). To keep things simple, in this book, IBM-compatible, Intel processor–based systems will be called PCs. To keep things even simpler, Apple Macintosh computers (powered by Motorola processors) will be referred to as Macs or Macintosh. The majority of presentations are created and delivered on these two platforms because they are the most widely used in general business applications and the most affordable systems on the market. In addition, they each offer a wide variety of powerful software and peripherals for creating presentations. Properly configured, both computer types make excellent multimedia platforms. In buyer preference surveys, Macintosh and PCs running Windows rank about even (see Figure 3-4).

The Mac

Because of its graphics-oriented pedigree and its compatibility with some of the most popular media editing tools available, the Macintosh, which runs on the processor made by Motorola, enjoys the loyalty of a large school of professional graphic designers, artists, producers, and developers. Until recently, Apple did not license its architecture to any other company; thus, fewer machines have been sold and less software has been developed. Nevertheless, hard-core Mac users, though a minority in the overall computer world, are known for their unshakable loyalty to the platform. Many professional multimedia developers tend to favor the Mac as the platform of choice. As of the end of 1993, the installed base of multimedia-capable Macintosh computers worldwide was approximately 3.3 million.

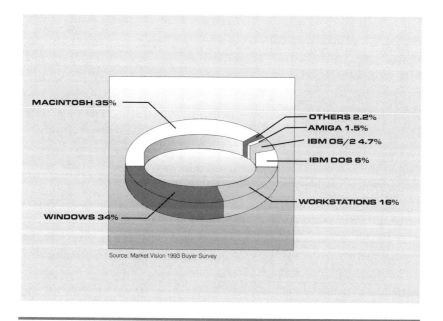

Figure 3-4 Buyer preferences for Multimedia products (source: Market Vision, 1993)

The PC

Personal computers that incorporate microprocessor chips made primarily by Intel are manufactured by dozens of companies, including IBM. They are sometimes called "DOS" or "Windows" machines, a reference to their operating system software. PCs command the marketplace by sheer force of numbers, outnumbering Macs by 10 to 1. But as of the end of 1993, only about 7.6 million out of the tens of millions of PCs installed worldwide were multimedia-capable.

The PowerPC and PowerMac

The relatively new machines built around the PowerPC chip from IBM, Apple, and Motorola hold a great deal of promise for multimedia. With the proper system software, the PowerPC is able to run standard Macintosh software and PC-compatible software as well as new software developed especially for the PowerPC chip. The PowerMacs (Apple computers built around the PowerPC chip) offer the same cross-platform performance in reverse. The chip, which is based on the RISC (reduced instruction set computing) architecture once found only on workstation-class machines, is built for speed and is particularly adept at handling large media files. As a result, most major multimedia vendors have announced plans to upgrade their products to optimize this new platform. By late 1994, a good selection of tools should be available for PowerPC-based systems, but you'll have to wait until at least early 1995 before the new platform is fully integrated into the multimedia presentation and delivery process.

Workstations

Computers based on powerful RISC processors are not typically considered personal computers. These high-powered workstations are costly and usually dedicated to a particular task, such as engineering or graphics processing. Few workstation vendors offer off-the-shelf multimedia configurations, but with the right components a workstation can be customized for multimedia development. Multimedia workstations are typically used in high-end installations for such projects as film animation, simulations, 3-D modeling, and virtual reality.

Operating Systems

The *operating system* refers to the programming (software) that turns a collection of chips and circuits (hardware) into a functioning computer. Both Macs and PCs are typically sold with standard operating systems. The Mac gained much of its prominence based on its intuitive graphical user interface (GUI). The interface refers both to what the computer user sees on the screen and to the command system that is used to run the system software. Most PCs run the MS-DOS operating system. MS-DOS by itself does not have an easy-to-use GUI, so most systems today are sold with Microsoft's operating system superstructure called Windows. With Windows running on top of MS-DOS, CPU performance suffers somewhat, but the IBM-compatibles become much easier to operate. A new version of Windows is due in early 1995 that promises to be much faster and more graphically oriented than Windows 3.1, which should prove positive for multimedia development.

As an alternative to MS-DOS, IBM introduced its own operating system, OS/2. OS/2 runs on many of the same machines as those that run MS-DOS. OS/2 sports a graphical interface similar in function to Macintosh and Windows. It currently has a much smaller market share than its main competitors, and, even though IBM has created a full line of multimedia development and presentation tools called Ultimedia, the tool set is limited and not widely used.

The UNIX operating system, which runs on RISC-based workstations, is familiar to many corporate computer users, particularly in the engineering and scientific fields. Its strengths lie in its raw power and its natural affinity for networking. While UNIX is a popular operating system for computation-intensive graphics work, it is not known as a general-purpose multimedia platform. But, as newer and faster communications technology allows multimedia presentations to be prepared and delivered over local area and wide area networks, UNIX is expected to play an increasingly important role in multimedia.

Compatibility

The Mac was built with graphics processing in mind and continues to include as standard equipment many of the components required for multimedia development and delivery. The standardized architecture of the Mac and the resulting standardization among third-party vendors makes presentations created on the platform relatively compatible with other Mac systems. This contrasts to the wide variations of compatibility between comparable PC products.

Unless specifically configured as multimedia machines, PCs do not include any of the components necessary for audio or video. For many machines, even graphics capability is an; afterthought. Numerous third-party vendors have rushed to fill the gap, but until relatively recently, few compatibility standards existed. The different specifications spawn incompatibilities that often make the development and delivery process on the PCs a challenge.

Within the last few years, a new standard has emerged to define the necessary components and specifications for a multimedia-capable PC. This standard, known as the *Multimedia Personal Computer* (MPC), makes it much easier to coordinate the specifications of the development and delivery systems. As shown in Table 3-1, the MPC standard specifies minimum configuration and performance for the hardware components needed for multimedia, including audio and graphics boards, hard drives and related connectors and processors, and CD-ROM drives. There are two MPC levels. Level 1 was created to include as many PCs as possible, but was quickly determined to be insufficient for most applications. Level 2 is considered the practical minimum for multimedia. Users who create presentations using the MPC standards can be reasonably certain that their presentations will play back correctly on a properly configured MPC-rated machine.

Layer Three: Storage Devices

The large files generated by graphics, audio, and video place extraordinary demands not only on the processors and RAM but on the storage drives and backup systems as well. Just one second of full-screen, full-motion, uncompressed video, for example, will occupy nearly 30MB. As with RAM mentioned earlier, having too much storage capacity is generally considered impossible.

Unlike the volatile memory of RAM chips (which store data temporarily during processing), hard drives, floppy disks, optical drives, and digital tape store data regardless of whether the computer is on or off. At least one storage device is a necessity for both development and delivery

	MPC LEVEL I	MPC LEVEL 2
CPU	386SX or compatible	486SX/25 or compatible
RAM	2MB	4MB
Magnetic storage	Floppy drive, hard drive	Floppy drive, 160MB hard drive
Optical storage	150K/sec CD-ROM with CD-DA outputs	300K/sec CD-ROM, CD-ROM XA ready, multisession capable
Audio	DAC/ADC, music synthesizer, analog mixing	16-bit DAC/ADC, music synthesizer, analog mixing
Video	VGA graphics adapter	640×480, 16-bit color
Input	101-key keyboard (or equivalent), two-button mouse	101-key keyboard (or equivalent), two-button mouse
I/O	Serial port, parallel port, MIDI I/O port, joystick port	Serial port, parallel port, MIDI I/O port, joystick port

Table 3-1   The MPC Specifications

systems. Most systems incorporate two or more types of storage. For further discussion of storage options, see Chapters 10 and 11.

Hard Disk Drives

The speed and capacity of hard disk drives makes them the most desirable development and delivery format for multimedia, but they are typically the most costly storage solution, particularly if more than one copy of the presentation is required. (See Table 11-2 in Chapter 11 for a cost/perfomance comparison.) Recently, a few hard drive manufacturers have introduced models designed specifically to handle the demands of digital audio and video (see Figure 3-5). These devices help eliminate the dropping of video frames and loss of synchronization that can sometimes plague even the fastest system.

Floppy Disk Drives

Floppy disk drives are limited to a maximum capacity of 1.44MB, but they have the advantages of low-cost hardware, widespread availability, and extremely portable disks. They can be used to load software applications and to back up and transport relatively small amounts of data from

Figure 3-5   Some hard disk drives, such as the Micropolis 3.5-inch and 5.25-inch AV drives, are designed specifically for the demands of working with audio and video

one system to another. A floppy disk drive is generally too slow to run a multimedia presentation, requiring that the data be first transferred to a hard drive.

Removable Magnetic Cartridge Drives and Removable Magneto-Optical Disc Drives

These storage devices have steadily improved their cost-to-capacity ratios in recent years. Although the faster models are capable of playing back multimedia, they are most commonly used to back up and transport large data files between systems.

CD-ROM Drives

CD-ROM drives, while far slower than hard drives, are a required standard for MPCs. The disks must be written on costly recording systems; but when used to widely distribute a presentation, the cost-per-megabyte is very low. Their large capacity is useful for storing and distributing software, clip media, or productions that demand high volume. CD-ROM disks and drives are capable of playing back multimedia presentations that have been optimized for the relatively slow playback rates. MPC Level 2 requires a double-speed CD-ROM. The newest models on the market feature triple and quadruple speeds.

Digital Audio Tape Drives

Digital Audio Tape (DAT) drives offer the most cost-effective means of storing data, but they are not randomly accessible and transfer data at a snail's pace compared to other devices. For this reason, DAT is used primarily as a backup storage medium.

Development and Delivery Storage

Development systems typically demand more storage capacity than delivery systems. When building the presentation, you will be creating numerous files, cutting, pasting, editing, and designing. The application software alone will occupy large amounts of storage space. While the actual size of your hard drive and other devices will depend on what software you use and how elaborate your presentations are, you should consider no less than a 300MB hard drive for the development system. More is advisable.

The hard disk capacity for the delivery system needs to be large enough to store the playback software and the multimedia presentation files. For example, you might need 500MB of hard disk while authoring your presentation, but the final file size of the finished product might be only 100MB. In that case, a delivery machine with the 160MB minimum specified by MPC Level 2 would be sufficient for playing back the presentation.

Layer Four: Graphics Boards and Monitors

The CPU and operating system go about their tasks invisibly within the confines of the computer cabinet. Before the user can actually see the GUI and the results of the CPU's computations, it is necessary to convert the digital data coming from the computer into analog signals that can be displayed on a computer monitor. More will be discussed about the differences between digital and analog signals in later chapters. For now, you should understand that special circuit boards, or cards, with digital to analog (DAC) chips are used to convert the output signal of the computer.

Graphics Boards

Most PCs and all Macs are sold with some form of graphics (video) board already installed, but dozens of options on all platforms are available for enhancing the speed and resolution of the display. Whether you go with the standard option or soup up the performance will be a matter of how much display power you need and how much you have to spend.

Graphics boards are differentiated by the number of *pixels* or dots they can display on a given monitor. The measurement is referred to as *video resolution* and is stated as the number of pixels displayed horizontally by the number of pixels displayed vertically (e.g., 640×480). Graphics boards are also differentiated by the number of colors they can display. Color resolution is measured in bits (e.g., 8-bit color). The bit measurements translate into a total displayable number of colors. Most graphics boards support at least 8-bit color, which translates to 256 different colors. Generally, more pixels equate to a larger, more detailed viewable image. The more colors a graphics board can display, the better it is at reproducing images. The 24-bit color boards, for example, are capable of displaying more than 16 million different colors, sufficient for reproducing photorealistic images. Video resolution and color resolution are discussed at length in Chapter 8.

The MPC Level 2 standard sets minimum resolutions of 640×480 and 16-bit color, but a 24-bit board is highly recommended. They can be purchased for as little as $500 and will give you the capability of displaying photorealistic images. Be aware, however, that the display capabilities of the graphics board have a number of ramifications that affect the other hardware that makes up the system. Greater color capacity requires more RAM and hard disk space as well as more processing. The 24-bit color boards run best on high-powered machines. If necessary, a *graphics accelerator board,* which takes over some of the graphics processing to free up the CPU, can be added to your system to speed the processing of large amounts of graphics information and improve display performance.

Monitors

Selecting the right monitor is one of the most important steps in building a multimedia system, yet it is often relegated to an afterthought. From the outside, most monitors look pretty much the same, but that is far from true. Monitors vary in image size, sharpness, and video frequency. A monitor that performs poorly will not only make the development process more difficult, it will make you less productive by causing eyestrain and fatigue. Simply put, the monitor is not the place to scrimp.

The easiest specification to evaluate is screen size. Like televisions, computer displays are measured diagonally. For multimedia work, the minimum practical size is about 13 or 14 inches. For better working conditions, a 15-inch or larger screen allows you to view the images you are creating at full size and still have the screen real estate to display media editing tools.

Monitors are also measured by the number of pixels they display. The ability to display a 640×480 pixel grid is considered a minimum standard for multimedia production. Most workstation–level monitors are capable of displaying more than twice that.

The sharpness of a monitor is measured by the dot pitch. Expressed in millimeters, *dot pitch* refers to the distance between pixels. The smaller the distance between pixels on the screen, the sharper the image will appear. For multimedia production, you should ideally invest in a monitor with a .28mm dot pitch; anything larger than .32mm should be considered unacceptable.

Most monitors in presentation work use cathode ray tubes (CRT), the same technology that drives television sets. Recently, liquid crystal display (LCD) monitors have entered the market. LCDs offer the advantages of being lighter and smaller, making them the ideal choice for laptop computer–based presentations. The latest active matrix LCD elements are found in projection technology that can be used to display multimedia presentations (see Figures 3-16, 3-17).

A few vendors are now marketing LCD monitors for desktop systems, (see Figure 3-6); but, in general, LCD technology has not yet advanced to the point where it can deliver the screen size, video resolution, color resolution, or response time necessary for a professional multimedia development system. Keep an eye on LCD monitors, though. They are becoming larger, faster, sharper, and brighter and could soon begin to challenge CRT displays.

Figure 3-6   The Radius LCD flat panel monitor display pivots for horizontal or vertical viewing and supports resolutions up to 1024×768

Layer Five: Sound and Video Hardware

This layer is the point at which your run-of-the-mill computer is transformed from a prosaic business productivity tool into a bona fide multimedia system. By adding the necessary hardware components, you and your machine step across the threshold from text and static graphics to animation, sound effects, narration, music, and video. The tools include graphics boards (discussed earlier), audio boards, video capture boards, and scanners.

Audio Boards

Sound is the one media type that almost everyone agrees qualifies a presentation as "multimedia." Sound engages one of the human senses that is completely ignored in presentations that use only static graphics and animation. With sound, presenters are given a new dimension to work in. It only takes a few experiments or demonstrations to realize that sound is one of the most powerful tools in the multimedia toolbox.

To process and output sound, a computer needs special audio processing circuitry that can convert the incoming signal from a sound source, such as a tape player, to a digital signal that can be stored and manipulated by the computer. In addition, the audio board (or card) is necessary to convert the digital sound information back to a signal that can be amplified and played through speakers. Both the development and the delivery system must have audio boards. To realize the full effect of audio as it has been captured and edited for the presentation, the delivery system requires a sound card capable of playing back the audio at the same quality level at which it was captured. (See Chapter 9 for a discussion of sound technology.)

Macintosh computers have always been equipped with 8-bit audio processors; the Macintosh AV line includes 16-bit digital audio capabilities as standard equipment. Because of its built-in capabilities, the Mac has relatively few add-in sound products compared to PCs. Basic PCs have no such built-in capabilities. Consequently, literally dozens of boards are available from more than 30 vendors. The boards range in price and performance from costly professional-level systems costing $600 or more to basic boards costing less than $100. They vary in the level of digital information they can process and the features they incorporate.

Most PC sound boards feature MIDI (Musical Instrument Digital Interface) as well as processing of digital audio. MIDI is a standard format that allows various types of musical devices to share digital information. Having MIDI capability in your computer gives you a range of creative options for audio. You can, for example, use any of the thousands of files of off-the-shelf digital music in your presentations. For the musically inclined, the addition of a MIDI keyboard or other input device allows the creation of original compositions. Macs do not typically offer MIDI support.

Once the sound is in the computer, a good set of computer speakers is needed to reproduce it faithfully (see Chapter 9). Again, a wide selection of manufacturers offers a range of products. In addition to the leading computer equipment vendors, many of the top speaker manufacturers have speakers systems for computers (see Figure 3-7). In general, speakers are not platform specific. Most will work with any type of computer or audio board that has the appropriate output connectors and can be used with either a development or delivery system, but it is important that the speakers' magnetic fields be "shielded" to prevent damage to monitors and hard disks.

Video Capture Boards

As with audio, video must be captured, processed, and converted to digital information from its original analog signal form in order for the computer to work with it. All video capture boards (or cards) use an analog to digital converter (ADC) to bring the video into the computer where it can be stored and edited. Some also provide digital to analog conversion so that video can be displayed on a monitor or recorded to videotape. The quality and performance the boards deliver vary dramatically from one product to the next. More than 100 video boards are on the market, ranging in price from less than $300 to more than $10,000. Within that range is a solution for virtually any application. Finding the right product is a matter of knowing what to expect from the board, what

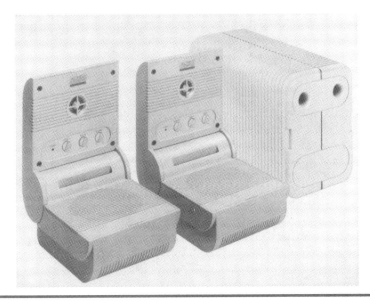

Figure 3-7   Self-powered speakers designed for multimedia applications, such as the Altec Lansing ACS 300, eliminate the need for a separate amplifier

quality level you need, and what the rest of your system will support (see Chapter 10).

Capturing video is a balancing act between getting enough video information to give you the quality you are after, without overwhelming the capacity of your system to support the huge volume of data. The better boards allow you to adjust your capture settings to optimize storage and playback for your system. The input parameters are generally determined by the capacity of your system and the intended form of output (videotape, hard disk, or CD-ROM).

While a video board is a necessity for capturing the video into the development system, the compression of the video data into usable file sizes can be done with software or hardware. Decompression of the video information for playback can also be accomplished with either hardware or software. Therefore, having a video board installed in the delivery system may not be necessary, unless the hardware compression system requires it. Software-only playback has the advantage of lower cost, but it has some severe limitations regarding the speed, quality, and image size of the video. Hardware compression (recording) and decompression (playback) solutions are available that deliver superior performance, but the cost of configuring both the development and delivery systems with video boards can be prohibitive, especially if you have several different delivery systems (see Chapter 10).

Dozens of good-quality boards are available for both the Mac and PCs. The top-end, broadcast-quality video capture, compression, and playback boards are currently found only on the Macintosh platform.

Video Cameras and Decks

Before you can begin integrating digital video into a multimedia presentation, you will need to capture the video images in an analog format, typically videotape.

Even if you have no video production experience, you are likely already familiar with many of the necessary video components from the consumer electronics marketplace. Many of the same camcorders and VCRs people use to capture the family picnic or play a rented movie can be drafted into multimedia production service. You should be aware, however, that quality varies widely from one format or product to another. The quality of the video sequences you digitize will correlate directly to the quality of the images you originally record; mediocre quality shows up more glaringly in the boardroom than in the living room. Also, if you decide to capture your own footage, make sure you are familiar with the basic techniques of video production (see Chapter 10).

VHS

The VHS video format is popular and the least expensive. Almost everyone owns a VHS-capable VCR, and hundreds of thousands of VHS cameras are in service. But most VHS equipment cannot capture colors with the accuracy and the resolution necessary for later capturing a quality digital image.

S-VHS

The S-VHS format, which produces higher resolution and better color fidelity than VHS, is being used by some companies to capture footage slated for digitization. Whether or not the format is adequate for your application is a matter of taste. Some dismiss S-VHS as inadequate for first-rate multimedia production work, but the price and versatility of the format make it worthy of consideration.

Hi8

The small size, reasonable prices, and high quality of the Hi8 format have made it a popular capture medium for multimedia applications. Hi8 cameras are even used by some news organizations to shoot footage that would be difficult to get with heavier and more costly professional formats.

Betacam SP

Currently, the Sony Betacam SP is a favored format among professional videographers in the corporate and television production environments. It delivers a high-quality image that is sharp and true to color. The professional-level performance of Betacam SP comes with a professional-level price tag. A complete system, including camera and playback deck, costs around $20,000 and up.

Editing Decks

Editing decks are available for all formats. They range from basic units to models with all the bells and whistles. For digital development purposes, much of the editing and special effects will be done once the footage has been captured digitally. Editing decks remain useful, however, for cutting large amounts of raw material down before beginning the memory-intensive task of capturing and compressing the video. As with video cameras and recorders, editing decks perform best in the hands of trained users.

Scanners

The images you use in a presentation can come from almost anywhere. They might be illustrations, photographs, forms, or documents. If they are not in a digital form when you acquire them, before you can use them in a presentation they require conversion with the use of a digital scanner.

Based on technology nearly identical to that used in photocopiers, scanners translate the tones and colors of an image into electrical values. Rather than using these values to charge a reproduction plate, scanners store the values as bits that correspond to a particular color or level of gray. The set of bits that represent the scanned image can then be processed with image editing software tools.

The first decision to make when considering a scanner is whether you need color or grayscale. Scanner technology has advanced to the point where some good color scanners cost less than $1,000, so unless you know for sure that you will never need color, a color scanner makes sense.

The most important specification for a scanner is its resolution, measured in dots per inch (dpi). Entry-level scanners can capture 300 dpi, but most offer 600 dpi. Higher quality scanners capture information at 1,200 dpi. The models used by top-quality print shops can scan several times that. A good 1,200-dpi scanner can be purchased for less than $1,500; the high-end equipment runs well over $20,000. Multimedia presentations are oriented to the computer screen; so for most presentation applications, a 300-dpi color scanner is adequate.

Scanners come in three basic formats: flatbed, slide, and hand-held (Figure 3-8). The flatbed type works much like a copier, in which the image is placed flat on a plate of glass. Many flatbed units offer optional rear-lighted hoods for scanning slides and transparencies. Slide scanners are configured to scan positive and negative films and transparencies. Hand-held scanners rely on the user to move the device across an image, making several passes if necessary. Optional guide tools can make the less

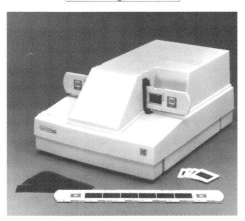

Figure 3-8   A few of the available scanner options: the HP IIcx
flatbed scanner, the Logitech PowerPage hand scanner,
and the Kodak RFS 2035 Plus slide and film scanner

expensive and less bulky hand scanners perform well, but their perform-ance cannot match that of a flatbed.

A few companies manufacture scanners that do not digitize an image, but rather capture it as an analog video signal. The signal can be fed to a video capture board for digitizing, or it can be overlaid on the computer screen using a special video overlay board. Video scanners are essentially video cameras set up on stands and are useful for grabbing images of three-dimensional objects that cannot be imaged using a flatbed or hand-held scanner.

Digital Cameras

Digital cameras are yet another way to bring still images into the digital environment for incorporation into a multimedia presentation. These devices function much like digital scanners, except the lens and digitizing circuitry are contained in a portable camera body. Digital cameras can feed a digital signal directly to a computer or they can store digital stills on disk drives or solid-state memory cards (see Figure 3-9).

Figure 3-9   Digital cameras, such as the Apple QuickTake 100 color digital camera, capture images without the need of conversion with a video capture card

Prices for digital still cameras start at less than $600 for a black-and-white model. For $750, the Apple QuickTake 100 provides simple capture of color images. Professional digital cameras can cost more than $10,000.

Layer Six: Media Creation and Editing Software

There are hundreds of software programs designed for creating, editing, and managing media. In every category you can find products for amateurs, professionals, and everyone in between. Many of the programs have overlapping functions as well. For example, some of the authoring and media integration packages discussed in the next section contain graphics, sound, and even video creation and editing tools.

Thanks to the growing body of clip media, you may find you have no need for media software at all. On the other hand, a multimedia presentation that does not need at least some custom media elements is rare.

Painting and Drawing

Painting and drawing utilities are included in the toolsets of most presentation authoring software, but the functionality of those tools varies greatly. Software programs created specifically for painting, image editing, or drawing are not designed solely for presentation work, but they can be very useful if the artistic demands of your presentation require the extra power.

In the hands of experts, the various tools for drawing and painting can be used to create everything from simple charts and logos to museum-quality masterpieces (see Figure 3-10). If so inclined, most people with a modicum of artistic talent can gain enough mastery of these programs to create visual images that are interesting and appealing. For a discussion of the types and uses of painting and illustration tools, see Chapter 8.

3-D Modeling and Rendering

3-D representations of real-world objects convey a sense of reality that simple 2-D images cannot. As a result, images rendered in three dimensions are often more interesting, engaging, and persuasive. The multidimensional dilemma for the presenter is that 3-D modeling and

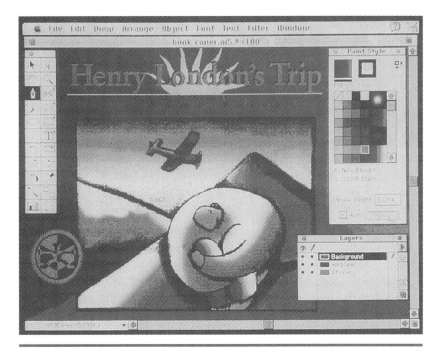

Figure 3-10 Dedicated drawing programs, such as Adobe
Illustrator, go far beyond the tools and functions
available in most presentation software packages

rendering programs are among the most difficult and demanding techniques in computer graphics. Most of the tools are designed for professional users and can be notoriously difficult to master. Recently, though,
this situation has begun to change. The newer programs come complete
with prefabricated models of common objects that can be manipulated
with relative ease. Online services now offer models of everything from
computer monitors to office furniture to buildings and airplanes.

The trick to selecting a 3-D program is to pick the one that is most
suited to your design needs. Some are meant for designing interior spaces
and architectural projects, while others are intended for the kind of
free-form, photorealistic look that virtual reality projects demand. Many
of the 3-D tools also include animation capabilities. See Chapter 8 for
further discussion of 3-D tools.

A few packages on the market have been designed specifically for modeling and animating fonts and logos. They use the same sophisticated rendering engines as the powerhouse tools but have easy-to-use interfaces that let you create impressive 3-D effects without the oppressive tutorial time (see Figure 3-11).

Animation

Most multimedia presentation authoring packages now provide at least some form of animation capability. Typically, they have automated functions such as screen transitions and bullet builds. A few include automatic chart animation and tools for directing an onscreen object to move along a user-defined path. Still, only a few of the media integration tools give you the kind of tools you will need to create a walking dog or a flying bird from scratch. If your needs go beyond what can be accom-

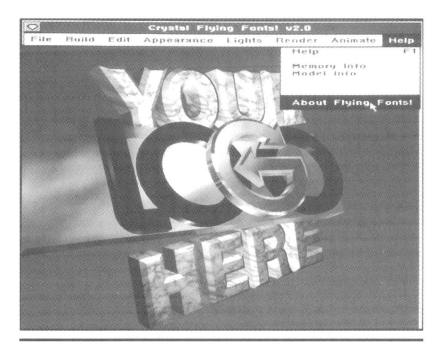

Figure 3-11   3-D logo design and animation packages, such as Crystal Graphics' Flying Fonts, give users of all skill levels the ability to create professional-looking graphics

plished in your authoring environment, you can look into one of the dedicated animation packages now on the market.

Dedicated animation packages range from easy-to-use, inexpensive software to professional-level tools with learning curves that are measured in months rather than days. The simpler packages automate many of the more difficult functions, but they will not give you the flexibility you might need if you have a background in animation or some experience with the traditional technique of cel-based animation.

Animation capabilities can also be found in many 3-D modeling and rendering programs as well as in multimedia integration and authoring tools. See Chapter 8 for more on animation tools.

Video Capture and Editing

Digital video capture and editing programs can be separate utilities or one in the same. They utilize the video capture hardware installed in the computer to capture and compress video, then edit it into a format that can be used by the multimedia authoring software.

Only a handful of programs are on the market, and more than half are designed for the Mac. Only Adobe Premiere is available for both platforms. Most of the better programs on both platforms can be used either to create an edit decision list (EDL) for offline analog video editing or to create digital movie files based on one of two video software extensions standards: QuickTime on the Mac or Video for Windows (AVI) on Windows-based machines. QuickTime and AVI, discussed in detail in Chapter 10, make it possible to work with video files the same way you would work with graphics, text, or any other digital file format (see Figure 3-12).

Once a digital movie has been created, edited, and stored, it can be played back on a different program, as long as the program understands the QuickTime or AVI format. That means the editing software used on the development system is mostly independent of the authoring program used for delivery. Movie files created on the Mac in QuickTime will play back on a PC system that is running QuickTime for Windows. Files created in AVI on a PC can be converted to QuickTime for running on a Mac.

The basic editing features in all the packages are similar, adhering to techniques common to analog video editing, but the degree of editing control varies and the ability to add titling and special effects is different from package to package. Prices range from about $200 to more than $1,200.

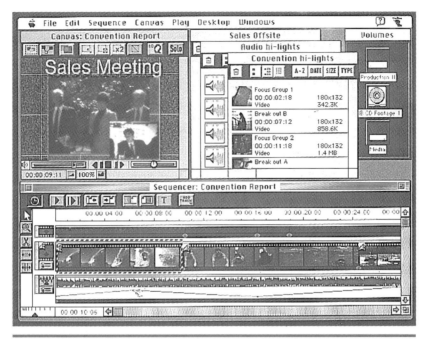

Figure 3-12   Video editing software such as Avid VideoShop uses many of the same conventions as traditional analog video editors

Special Effects

Many of the video editing packages provide at least a few special effects and transitions. But for those users who have mastered the basics and wish to go beyond the mundane, a handful of tools are available that let you manipulate both still and video images. Most of the tools in this category are oriented toward video or animation production. Some of the special effects products, which can range in price from $99 to $1,000, may be overkill for all but the most daring designers. Others are simple to use and provide maximum impact with minimum effort.

The two most popular special effects techniques are morphing and warping. *Morphing* means transforming one object into another (see Figure 3-13). Morphing provides an entertaining way to show objects in relationship to one another. *Warping* involves distorting the shape of an object, and is often used for comic effect or to add a surrealistic feeling to a presentation.

Figure 3-13   Morphing packages, such as Gryphon Software Morph, create animated (movie) sequences depicting transformation of one object into another

Sound Capture and Editing

Sound editing software—programs designed to adjust the length, quality, and special effects of sound files—range from the simple utilities provided with the Windows multimedia extensions to professional-level editing tools. Basic sound editing utilities can also be found in some authoring and video editing software packages. Most sound boards come bundled with a sound editing utility.

How much sound editing power you will need depends on what types of sound you will be using. All sound editors perform the basics. Special effects, such as fade-ins and fade-outs, echoes, reverbs, and stereo panning, are available with the more advanced tools. *Looping* (setting up a sound file to repeat) can be useful as well, especially for continuous playback while your audience files in or out of the room. Some or all of these features can be found on sound editing packages that range in price from $70 to $700.

An important feature to watch for when choosing a sound editor is the type of files it supports. Macintosh computers use different file formats than PCs, although the AIFF (AIF) file format is cross-platform. The most popular file format for PCs is the Wave (WAV) format. If your presentation plans include playback on both platforms, then your best bet is to invest in AIFF-compatible software.

Music and MIDI

For business presenters with a musical bent, MIDI sequencing software may be just the ticket to enliven an otherwise utilitarian toolset. MIDI sequencing software permits you to integrate a synthesizer into your hardware system, opening up a brave new world of sound and music capabilities. Many of the sound boards for PCs include a basic MIDI synthesizer.

While Wave and AIFF files are complete digital representations of the analog audio source, MIDI files consist of a set of instructions that are executed by the MIDI sound board. A MIDI file instructs the board to play a certain note on a particular instrument for a specified duration. The board itself creates the sound and outputs it to the computer's speakers (see Chapter 9 for more on MIDI).

While music and sound can add a new dimension to multimedia presentations, the technical and musical skills required to create original music with these tools is considerable. To make the process easier, thousands of prepared MIDI files are available from online services, clip media companies, and shareware libraries. These files can be played back by any MIDI-compatible synthesizer or board and provide a method for including a wide variety of musical styles.

Media Management Tools

The process of designing and building a multimedia presentation can require dozens, if not hundreds, of media elements: graphics, text, video, audio files. Managing the files in all their various formats can be a daunting challenge. Indeed, effectively organizing, sorting, storing, and retrieving media elements is crucial to realizing the productivity gains promised by digital media. Software publishers offer a number of tools that have been created specifically to act as librarians and researchers.

Most of the products categorize files by type, name, and location, or by reference information you attach to each file. They perform search and retrieval functions similar to many popular database programs. Typically, they provide a thumbnail view of the file's visual identification before it is opened, copied, or transferred.

Some of the better tools will even open up a file directly in the program that created it in the first place. A few are designed to operate over a network, providing the entire presentation development team with organized access to a set of files.

Layer Seven: Authoring and Media Integration Software

You have a choice among dozens of programs designed to build a multimedia presentation. The packages are loosely referred to as presentation software, presentation graphics software, media integration software, or authoring software. As explained in Chapter 5, where the products and categories are discussed in detail, all of the packages can generally be described as authoring tools (a term sometimes erroneously reserved for professional multimedia development software).

As with most of the tools, the package you select will depend on the type of presentations you plan to create and the level of product sophistication you require. Some software packages give you the barest essentials, while others provide more functionality than the average user could employ in a lifetime of authoring.

Many of the general-purpose presentation authoring programs are available both on the Macintosh and for Windows. In many cases, a presentation created in one program on one platform will run in the same program on another platform. A presentation will not run on different authoring software, though some products allow you to incorporate a presentation authored in one package into a different program.

For presentation delivery, most of the programs offer playback-only versions of the software with the main program. With these *run-time* utilities, you can equip a delivery system to play the presentation without having to load the entire authoring program onto the system.

Authoring tools can be grouped into three basic tiers: multimedia-capable presentation software, dedicated media integration software, and professional multimedia development software.

Multimedia-Capable Presentation Software

Programs that allow you to create a series of slide-like images and run a slideshow-like presentation from the computer have been around for more than seven years. These presentation graphics packages are still popular for creating images for output to 35mm film, transparencies, and color handouts. *Screenshow* (computer-delivered) presentations are growing in popularity, and most of the packages now have at least basic multimedia capability.

Most packages can add media elements such as sound clips, videos, or animations directly to a screenshow presentation as easily as importing a chart or graph. In addition, most feature some form of painting and drawing tools, image transitions, outliners, and slide sorters for putting together and playing back an entire presentation in one environment. A few allow you to animate objects or create animated cartoons. Many have also added basic interactivity features that allow presentations to jump and branch to other slides and topics. Finally, some packages now feature timeline views that allow precise control of the timing and synchronization of events on screen.

Dedicated Media Integration Software

Where multimedia-capable presentation software essentially has added multimedia after the fact, a relatively new generation of presentation software has appeared in the last two or three years designed specifically for multimedia presentations. As with the previous category, you will not find media creation tools such as video editors, sound editors, or capture software in most of these products. Rather, they provide a framework for consolidating many different media elements into a single presentation. It should be noted here that the dividing line between these and the previously mentioned packages is hazy. Many of the tools and functions overlap; this is particularly true of products that have been optimized for onscreen multimedia presentations but include all of the more traditional slidemaking functions as well.

Even though media integration packages offer more sophisticated tools, they are relatively easy to use. In general, they do not offer the charting, graphing, painting, drawing, or outliners found in multimedia-capable presentation software. But several of the tools in this class are well designed for creating animations and establishing interactive links.

Professional Multimedia Development Software

Complex multimedia presentations—any situation in which the creator of the program wishes to script complex actions and connections—often require more power than is provided by slideshow presentation and media integration tools. Professional development tools, which typically cost $700 or more, are designed to create various levels of interactivity, including the ability to capture or read data from the user/audience and customize actions based on that information.

Packages in this tier go well beyond media integration tools by using scripting languages to extend the software's capabilities. The earliest versions of high-powered authoring tools were difficult to master. Current versions, while easier, still demand a great deal of study and practice.

Like media integration tools, professional authoring systems utilize the familiar metaphors of screens, scripts, and timelines. A few of the tools in this category use an icon or object-oriented metaphor to ease the development process by prepackaging certain programming operations. All of these tools require a relatively sophisticated knowledge of programming techniques and principles, but in the right hands they open up nearly limitless possibilities for multimedia presentation.

Layer Eight: Delivery and Display Systems

The final tier of the multimedia system includes the playback and transport devices as well as the display technology used to make the presentation visible and audible. The delivery and display choices you make affect virtually every other component in the system. Delivery and display, including the products and the presentation environment, should be given ample consideration early in the system configuration process.

Delivery Hardware

The storage systems discussed earlier in this chapter encompassed all of the devices that can be used for transporting, distributing, and delivering your multimedia presentation. Table 3-2 summarizes the strengths and weaknesses of the various options. For a more complete discussion of fixed and transportable drives, see Chapter 11.

Display Options

The final step for any multimedia presentation is putting it on the screen for an audience to see and hear. The type of presentation and the size of the audience will determine what display system is right for a particular presentation and a particular audience. Display technology

MEDIUM	CAPACITY	TRANSPORTABILITY	MEDIA COST	PERFORMANCE
Floppy	Low	Excellent	Low	Slow
Hard Drive	High	Poor	High	Fast
Disc cartridge	Moderate	Excellent	Moderate	Moderate
Optical	Moderate	Excellent	Moderate	Moderate
CD-ROM	High	Excellent	Low	Moderate
Tape	High	Excellent	Low	Ultra-slow

Table 3-2 Delivery and Storage Systems

Case Study:

Company: InVision Technologies, Inc., Foster City, California

Business: Manufacturer of an explosives detection system

Objective: Eliminate the need to ship the company's large and complex security product to cities around the world for demonstration, by using a multimedia presentation to explain the system and simulate its functions

Some products have to be seen to be appreciated. Take, for example, InVision Technologies' advanced explosives detection system. Known as the CTX 5000, the airport security system employs CAT-scan technology and 3-D rendering to electronically "unpack" luggage and let the operator know if explosives or suspicious metals are lurking inside.

According to product applications specialist Tom Meyer, many of InVision's aviation security clients are unfamiliar with advanced computer technology. As a result, they have a difficult time comprehending how the system works. Once they see

it in action, he says, they immediately understand the product and its value to their security operation. InVision's only problem is that the CTX 5000 weighs 6,000 pounds and the company's clients are scattered all over the world, making on-site demos a logistical nightmare.

To help clients envision how the CTX 5000 works without actually lugging the product along, the company turned to multimedia. Back in 1992, during InVision's initial foray into new media technology, it went to an outside service for help. San Francisco–based interactive multimedia developer Haukom Associates created a multimedia simulator using a Macintosh IIfx equipped with a RasterOps 24STV videographics board and a laserdisc player. The presentation, which was displayed on an overhead projector via an LCD panel, depicted an airport guard walking up to the CTX 5000, turning it on, and loading in baggage. As a simulated bomb threat situation developed, the screen switched to the graphical interface of the CTX 5000 and demonstrated how the machine detected various explosives.

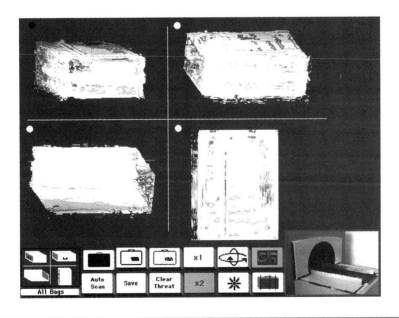

InVision Technologies

While this presentation was extremely effective for InVision, the combination of the Mac, laserdisc player, overhead projector, and LCD panel was still not as portable as it needed to be. With multimedia technology becoming more compact and easier to use, InVision began creating its own multimedia presentations and delivering them from a lightweight Apple PowerBook notebook computer.

For a presentation to Brussels airport officials, which resulted in the company's first sale, Meyer and his associates created a four-part program. The program begins with an introduction to the company, created by Meyer with Aldus Persuasion authorizing software. The first section is a simple electronic screenshow made up of text, scanned photographs, and graphics. The presentation then transitions to an animated segment created in Macromedia Director, which serves the same function as the previous simulator project. It begins with video that was recorded in Sony Betacam SP and digitized to QuickTime format. Shot from the point of view of the baggage, the video shows the bag entering the CTX 5000 on a conveyor belt. Ahead of the bag is the spinning gantry, which has the X-ray tube attached to it. The video then fades into an animation that illustrates how the machine X-rays the baggage and sends the image data to the computer to be analyzed and interpreted. Immediately following the animated sequence, the presentation switches to a training program created in a Macintosh package called Runtime. The program depicts the CTX 5000 from the operator's viewpoint, complete with the monitors and button controls. In this way, the audience sees what the operator would see and do while using the detection system.

After the animated sequences, the presentation switches back into Aldus Persuasion for a discussion of tests conducted with the CTX 5000, as well as details about its use in various airports around the world and specific plans for its implementation at the Brussels airport. The presentation again transitions into an animated program, this time created in a Macintosh program designed for manufacturing called Extend. This animation shows how the CTX 5000 would be set up in the Brussels airport and explains how it would speed up baggage handling. To conclude, the presentation goes back to Persuasion for some summary slides.

"The people in Brussels said that it was the best presentation they had ever seen," reports Meyer.

and the considerations involved in selecting the right tools are discussed at several relevant points in this book. Refer to Chapters 11 and 14.

The following is a rundown of the key display technologies. In all cases, the display devices can be used with any of the computer platforms, as long as the necessary connectors and input/output signals are compatible.

The one thing to keep in mind about display technology is this: quality counts.

Presentation Monitors

When the audience is too large or far away to allow people to view a presentation directly from the computer monitor, a presentation monitor offers the convenience of direct view and good image quality. Large-format monitors come in sizes from about 27 inches up to 42 inches (see Figure 3-14). Prices start at less than $1,000 for a common television set and go up with the screen size to as much as $13,000. Resolution ranges from basic video (NTSC) up to workstation quality. Some monitors display only analog video signals. Multisync monitors are capable of handling video as well as a range of computer (RGB) signals.

CRT Projectors

For large-scale presentations, CRT projectors offer bright, large images and compatibility with a wide frequency range (see Figure 3-15). They tend to be relatively heavy and impractical for traveling presentations. Prices range from about $5,000 up to $50,000 or more.

Figure 3-14   Panasonic FT-2700 27-inch presentation monitor

Figure 3-15   Sony VPH-1271Q CRT projector

LCD Projectors

These relatively new display devices use LCD elements rather than internal CRTs to project images. The top-of-the-line models feature 1024×768 resolution, multiple signal handling, and good full-motion video display. Their primary appeal rests in their light weight and ease of operation (see Figure 3-16). Prices range from less than $8,000 to more than $20,000.

LCD Projection Panels

These systems sit atop overhead projectors to display data and video signals from the computer. They are a valuable tool for their portability, but they cost in excess of $4,000 for high-resolution, full-color models. Properly equipped LCD panels can also project full-motion video. A few companies have LCD projection panels that, when a backlight is added, double as LCD direct-view monitors (see Figure 3-17).

Videowalls

Popular in trade shows in malls and other public venues, videowalls utilize video projection cubes or CRT monitors in a grid or other pattern

Figure 3-16   Proxima 8300 LCD projector

Figure 3-17   Sharp 1650 LCD overhead projection panel

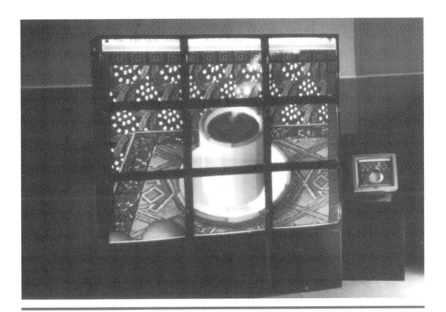

Figure 3-18 RGB Spectrum MediaWall

to create large and eye-catching displays (see Figure 3-18). They serve best for large audiences and attention-getting presentations. Videowalls usually require trained specialists for installation, programming, and operation. Prices range from just under $20,000 to more than $100,000 depending on the type and configuration.

CHAPTER 4

Few things are brought to a successful issue by impetuous desire, but most by calm and prudent forethought.

—Thucydides

DESTINATION MULTIMEDIA: PLANNING THE PRESENTATION

With multimedia, you feel a nearly irresistible temptation to jump right in, start up the engine, and pull away from the curb with a trunk full of text, graphics, sound, and video. This is understandable: the thrill is in seeing and hearing how all the media elements work together. But, before you pack your bags, you must know where you are headed. Before you depart, you must plan your trip.

The Multimedia Process

Would you set out on a six-city business trip without first planning your itinerary, your stops, your timing, and your transportation? The multimedia production process for creating business presentations can take many routes. You can select from multiple media types, dozens of tools, and hundreds of approaches. By definition, multimedia gives you more destination choices, options, and possibilities than traditional forms of presentation.

Ironically, the more freedom you have, the more planning is required. You must know where you are going if you are to going to get there, get there safely, and get there in style. Likewise, the more important and elaborate your presentation, the more thorough your preparation must be.

A useful rule of thumb: By the time you boot your authoring software, you should be halfway there. In other words, at least half of the work in a well-designed multimedia business presentation occurs before you ever start using the hardware and software.

The multimedia design and planning process consists of four basic steps:

- Analyzing your *audience*
- Determining your *goals* and *objectives*
- Accounting for *logistics*
- Planning your *production*

Understanding the process will help you select the right hardware and software, meet your audience's needs, and achieve your presentation goals (see Figure 4-1).

Asking the Right Questions

Good multimedia planning, as with most things, begins by asking the right questions. These may be questions you ask yourself or questions you pose to others. You might have the answers at hand or you may have to do some digging. If you force yourself to step through the questioning process, avoiding the tendency to jump ahead, you will discover at the end

Figure 4-1   Plan your excursion into multimedia before
you embark

that all the disparate elements fit neatly together. The presentation will
have logic, intelligence, and harmony. You will be more effective as a
presenter or producer, and you will experience fewer unpleasant surprises.

In a sense, as you begin the design process you are like a reporter
assigned to a story. Journalists are trained to get the essential facts by
ferreting out the answers to six basic questions: *Who, What, When, Where,
Why,* and *How* (see Figure 4-2). No writing starts until the questions have
been answered. The more complete the answers, the better the story. The
questions do not have to be answered in order. But you will notice in this
chapter that the answers to each question have a bearing on all of the
others, so you should be as comprehensive as possible.

WHO Is the Audience?

Virtually every presentation is designed to bring an audience into
accord with the presenter's message. As explained in Chapter 2, even if
the overt goal is to assist the audience with a better understanding of the

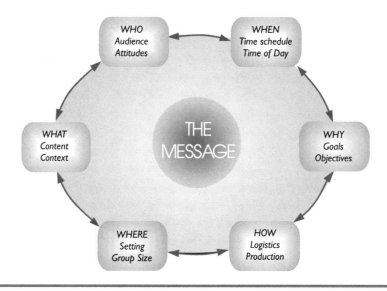

Figure 4-2 All planning considerations revolve around
the message

information, an effort at persuasion is almost always behind the scenes. A presenter is always selling something to someone—a product, an idea, a concept, an opinion, a point of view—and as in any sales situation, the most important factor for success is knowing the customer. In the case of business presentations, the audience *is* your customer, regardless of your relationship to them.

To be effective, you must understand as much as possible about your audience's background: their responsibilities, their demographics, and their expectations for the presentation. What are their hot buttons—what will excite and persuade them? What are the hidden or unspoken rules that your audience expects you to understand? How will their job responsibilities affect your approach?

You can categorize or characterize an audience in several ways. For purposes of this discussion, they will be identified as decision makers, specifiers, and influencers. You do not need to view audiences in these categories, but you should try to determine their overall character before you delve into specifics. The approach or attitude they bring to the presentation will have a direct bearing on the multimedia technology and techniques you employ.

Decision Makers

Decision makers will typically be in attendance to gain an overall feeling or impression of you as the presenter, your organization, and the value of the information or solution you are offering. Their hot buttons are big-picture, bottom-line considerations, such as return on investment and future growth. They will be heavily influenced by the depth of information and your aesthetic execution.

Remember, decision makers often approach a presentation with preconceived notions. If you suspect they are harboring prejudices or negative opinions, for example, counter those impressions by demonstrating your current strengths. Did your product have trouble with reliability last year? Demonstrate the improvements with animations, diagrams, or customer testimonials. Use charts, graphs, case studies, testimonials, and other visuals to convey the key attributes that concern the decision makers: reputation, financial stability, reliability, productivity, and so on.

Decision makers tend to have limited time for attending presentations, so keep it brief. Find out in advance how much time you have been allocated and stick to the schedule.

Specifiers

Specifiers are individuals or groups charged with making "specific" recommendations. They may or may not make the final selection or determination. They typically have a preexisting knowledge base on the subject at hand, so assessing their depth of experience and training is critical before crafting the presentation. Failing that opportunity, you should design an interactive presentation that contains multiple levels of information. That way, as the presentation proceeds, you can adjust and access the degree of detail and expertise suitable to the specifier.

As you will learn in greater detail in Chapter 6, interactive multimedia powerfully addresses the needs of an audience with unknown skill levels. Using presentation authoring software with the appropriate interactive features, you can not only drill down to deeper and deeper levels of detail, you can also jump randomly to the areas of greatest interest, skipping material that the audience already knows. Specifiers will tend to be less impressed by flash and sparkle. They may suspect that your use of sound and visuals are an attempt to cover up for lack of substance or accuracy. Keep the visuals clear and to the point.

Influencers

Influencers are those individuals in the audience who do not have the authority to make specific recommendations or decisions but can, if converted by the message of the presentation, have a direct effect on your success or failure. They may have come in search of facts, or they may have only a general interest in the subject and are looking for ideas to take away with them. This group will respond as much to the energy and enthusiasm of the presentation as to the information it contains. They are typically the audience for evangelistic presentations, in which the goal is more to engender an underlying sense of excitement and drama than to impart specific information. The highly entertaining aspects of multimedia presentation—humorous animation, lively video clips, music, and dramatic graphics—are ideal for reaching an audience of influencers.

Occupational Factors

Occupation is one of the more obvious factors that must be considered as you design your presentation strategy and select your media. Whether your audience is composed of people who serve in the same job capacity or mixed professions, you should be aware of how individuals with different jobs tend to react to media. The following four occupational groups—scientific and engineering, architectural and industrial design, sales and marketing, and manufacturing—are used as examples of how responses to your presentation will vary. These are broad generalizations, not meant as hard and fast rules, but rather as demonstrations of the audience analyzation process.

Scientific and Engineering Professionals

This group tends to be influenced by presentations grounded in facts, accuracy, and logic. The more precise and demanding the occupation, the more the information must be supported with research data, clinical studies, and careful citation of sources. Hyperbole has no place here.

Complex and difficult data often are best explained with 2-D or 3-D charts and graphs to illustrate relationships. Various types of charts and graphs are used to differentiate information, show trends and correlation, and suggest averages. Multimedia techniques allow the charts to grow and move as desired for further emphasis. Using a charting program in real time, a presenter can instantly change from one chart type to another using the same data. Clinical case studies illustrated with scanned photographs and video sequences add further detail by including actual people, products, places, and events.

Architectural and Industrial Design Professionals

These people (e.g., interior designers, fashion designers, landscape designers, and industrial or product designers) tend to respond very favorably to artistically rendered visuals. As artists themselves they are highly appreciative—but also highly critical—of the work of others. The use of scanning technology as well as video capture can generate both ideas and context. Many of these professionals already use sophisticated software to do mock-ups and renderings as well as make-overs and "what-if" scenarios. When presenting to this audience, you will want to be certain that your images are of the highest artistic and design merit. Avoid anything tacky or shoddy. Packaged clip art rarely satisfies members of this group. An original and stylish logo, for example, will demonstrate your commitment to excellence (see Figure 4-3).

Sales and Marketing Professionals

This is a broad-ranging group, but they can generally be said to be engaged in the art of persuasion as a livelihood. They tend to be concerned with such issues as market trends, new product developments, sales results, and market opportunities. Because selling is a highly interpersonal profession, these individuals tend to be affected by human touches in presentation—humor, anecdotes, stories—anything that brings abstract information down to a personal level.

They will respond positively to fast-paced, upbeat presentations. The multimedia design model for this group is generally the same as that used by advertisers. Be entertaining, informative, and go for a subconscious connection. Music and sound effects are usually well received when they help to create a mood. These individuals might not be trained in graphic arts or photography, but they will catch inaccuracies in specific areas.

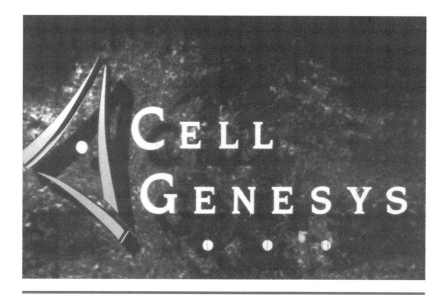

Figure 4-3   Design professionals will expect excellence in design, as in this 3-D logo created with Pixar Typestry

They might notice, for instance, if you make color or design errors when incorporating a familiar brand name design or logo.

This group also enjoys product shots, particularly new products or renderings of future products. Video of consumer/customer reactions or experts in their field offering specific solutions work extremely well here. Finally, this group enjoys and appreciates interaction with the presenter. Design a presentation to solicit their participation. If possible, let their feedback influence the direction of the presentation. Interactive multimedia techniques, for example, allow you to take votes or count responses, enter them into the live presentation, and instantly generate animated charts.

Manufacturing Professionals

Whether they are line managers or project coordinators, manufacturing professionals are concerned with materials, processes, labor issues, and supply questions. Manufacturing processes are traditionally documented through process diagrams. Multimedia enables animation to quickly illustrate the steps of a procedure over time. Video documentation of procedures can be generated in real time and captured for playback in later presentations. Labor issues involve people—everything from safety to compensation. Photographs of safe, happy employees using a product or service will be welcome and can provide a subtle but effective backdrop to the point you are trying to make. Supply issues, productivity issues, and waste concerns tend to come down to statistical analysis. Multimedia can help manufacturing professionals to visualize data in charts. Photographs let them see products, machines, and environments.

Demographics

Obviously, knowing in advance the age, gender, cultural orientation, income level, and education level of your audience is necessary for targeting your presentation. All these elements will also help to determine the types and degree of multimedia you adopt. Again, the rules are not chiseled in stone, but here are some general guidelines to follow.

Note: For the most part, these guidelines will apply to all presentation types, including electronic brochures, training, and kiosks; but for purposes of clarity the focus here will be on in-person presentations with multimedia used as speaker support.

Age

The average age of your audience should be taken into account when deciding on the type of sounds and visuals you use. Each generation responds to the imagery of its time. Musical tastes, artistic styles, and values change in a society over time. Taking these factors into account in the media you choose will help you communicate your message with more impact. Younger audiences, for example, respond positively to rapidly changing images and a driving sound track that are in tune with current trends in art, music, and culture. For this group, use music videos and television advertising as your models. Wherever possible, and within the limits of copyright usage, try to incorporate images of popular celebrities, products, and events. Like most audiences, younger audiences also appreciate humor, but typically it must have a sharp edge. Self-effacing humor works well. So does a bit of sarcasm or satire.

Gender

If your audience is predominantly male or female, you will be well advised to be cautious about creating a presentation that is stereotypically aimed at either. Assumptions about what appeals to men or women are highly subjective and prone to misjudgment. You are on safer ground appealing to the sensibilities of the audience, rather than gender stereotypes.

Socioeconomic Background

Cultural background, education level, and economic status will determine not only the tone and depth of information, but also the images, music, and narration you choose. Again, you do not have any specific rules to follow, except to be highly sensitive to media elements that might offend or confuse. Do not attempt to target a specific cultural group, for example, if you are not completely confident that you understand the effect the material will have. On the other hand, multimedia gives you the capability to flavor your presentation with a variety of cultural references. You are free to experiment with images and sounds that evoke certain feelings or associations. Because the material is extremely flexible, you can test it and make alterations as often as you wish until you get it right. The same procedure applies to adjusting your presentation to education levels and economic status. Experimentation will tell you if you are talking over the heads of your audience. It will also help you to avoid the mistake of talking down to them.

WHAT Will Be Presented and WHY?

Even before analyzing the target audience, you should have a clear idea of *what* material will be included in the presentation and *why* you are presenting with multimedia in the first place. What is it about the message that requires the extra time and effort involved in creating multimedia? More simply put: "Why bother?"

Almost all business information that passes from place to place, person to person, or company to company receives some form of presentation or organizational enhancement. Most written documents have been organized into sections and paragraphs in someone's idea of a logical structure. Books, newsletters, and magazines go further with graphical elements and type treatments. Even messages transmitted over the phone are enhanced by the speaker with vocal inflections and emphasis.

Presenters using no technological support at all make hundreds of small enhancements designed to better communicate the message. The process of deciding what enhancement is necessary hinges on three points:

> The *intent* of the message

> The level of *importance*

> The degree of *complexity*

It stands to reason that virtually any content could benefit from multimedia enhancement. A salesperson showing a new product in action is far more effective than merely calling prospective customers and *telling* them how great the product is. Presenting a colorful and animated project management diagram can be more powerful than writing a detailed prose description.

Nevertheless, the level of presentation must be kept in scale with the intent of the message. Technological overkill can seriously detract from the message or even bury it. A simple project update for a group of managers, for example, might be intended more as a pep talk than a working and planning session; in which case, the personal sincerity and enthusiasm of the presenter is of primary importance—using electronic charts, animations, sound effects, and such would be like using a CAT-scan for a toothache. See Chapters 7 through 10 for more information on the various media types and how to use them effectively.

Similarly, if the information does not lend itself to visual or auditory enhancement, an audience might become confused or misled. Explaining in as many words that management *felt* it was time to pull the plug on a project because it was no longer in keeping with the direction of the company would be far cleaner and easier than generating a multimedia presentation on the subject—the cost and effort of the presentation would be contradictory to the message itself.

Improperly applied multimedia enhancement can also cause information of relatively small importance to be misperceived as a top priority or central focus. Interim sales figures, for example, may mean little in relation to the year's overall performance; but if plugged into the same presentation template used for year-end numbers, they may become a priority item in the mind of the audience.

Presentation overkill most often occurs when the presenter is anxious about speaking to the audience, is trying to divert attention from the real issue, or lacks confidence in the content. Never let your use of multimedia become a fancy border around a dull or out-of-focus picture.

Finally, and this need hardly be said, if the message comes through loud and clear with a few simple words, it has no need at all for enhancement. Multimedia presentation is at its best when helping to clarify difficult or complex messages. It should be used to cut through layers of misunderstanding, knowledge gaps, and lack of experience, but not to gild the lily.

Goals and Objectives

Ask yourself, what are you trying to accomplish? What are your goals? What are your objectives?

Goals refer to what you want your audience to do, think, feel, or know when the last image fades to black and the music dies away. Should they take a particular course of action? Develop a new opinion? Be convinced about a certain project, idea, or concept?

Your *objectives* involve the specific approach, tactics, programs, and procedures that you will use in the process of reaching your goals. Will your multimedia presentation be designed to teach, to train, to inform, to sell, to entertain? Most likely it will be one or a combination of those objectives.

Good goal statements for your presentation should include measurable results whenever possible. The following is a basic goal statement for a hypothetical presentation designed to introduce new pollution control measures to an audience of decision makers:

Environmental Policy Presentation — February 22
Primary Goals

- Educate management on the problems.
- Instruct them on the proposals for the new pollution control policies.
- Create positive feeling about the proposals.
- Convince management that the project should go forward.
- Obtain executive sign-off by May 1.

In this example, notice that goals become increasingly more concrete and measurable. Goals focus on audience reactions and motivation. Specific measurable events (the date for sign-off) are included.

Specific objectives for multimedia design of the same proposal would look something like Figure 4-4.

Content

From the statement of your goals and objectives will follow the content and concept. The *content* or subject of your presentation defines the specific facts, figures, and other information that you plan to communicate. The *concept* involves how you will present your content in order to meet your goal. For example, to present ideas on new products, you may choose to include surrealistic visual renderings with emphasis placed on the look and feel and not the details. This is regarded as a "high concept" approach. You are relying on a nonrational communication of ideas rather than a strictly logical or literal approach. High concept is

POLLUTION CONTROL PRESENTATION—FEBRUARY 22

Objectives	Media Selection
Add *context* to the message. Set the mood.	*Dramatic music. Video and still images* of effects of pollution on environment.
Outline pollution problems and potential solutions.	*Animation* showing pollution process. *Photo* of results. Graph showing costs of cleanup and fines.
Introduce details of the cleanup plan and new policy.	*Chart* outlining five steps to cleaner environment and better policy.
Drive home the long-term cost and corporate image benefits.	*Charts and graphs.* Example using live spreadsheet.
Convince them that the plan is the right course of action.	*Video images* of corporate logo overlaid on blue sky, clean water. Energizing *music*.

Figure 4-4   An example of presentation objectives

useful when you need to prove to your audience that you and your organization are stylish, progressive, and intelligent.

Using role-playing videos, to demonstrate how new products are being received by the public, is another example of a high concept approach. Which is not to say that multimedia presentations are always high concept. In most courtrooms, for example, overly dramatic presentation is generally counterproductive. In fact, studies have proved that juries are better persuaded with graphics that use simple line drawings and outline shapes rather than realistic 3-D renderings.

WHEN Will It Be Presented?

You may not always know the time and place of your presentation. Will it be in the early morning before the coffee kicks in; mid-morning when they will be sharp, alert, and eager; after lunch when they will feel sluggish; or in the evening when they may be sleepy? The presentation guidelines for time of day are well known and commonsensical. Studies show, for example, that energy levels go down directly after large meals, such as lunch and dinner. If you are the après-meal presenter, you are advised to pull out all the stops with light, sound, energy, and humor.

Entertainment value in a presentation is welcome most any time of day, but it can be critical to success when you come up against an exhausted audience—not to be confused with a disinterested audience. Multimedia techniques not only give you a far wider range of entertaining options than traditional presentation techniques, they also let you adjust the elements for the time of day. Your standard presentation, for example, might open with classical guitar music, but you could have a rousing Sousa march on the disk in case you are given a late-afternoon time slot.

The time of day also has an effect on *length*. In general, attention spans will diminish as the day wears on. When possible, you should keep your material under an hour. Give the audience an agenda that includes breaks so they can gauge the time. Never ask an audience to sit longer than 90 minutes without a break. Multimedia presentations let you adjust your timing by compartmentalizing information, in effect, creating subset presentations of differing lengths. The principles are the same as for designing a presentation that will be delivered to different audiences at different times. Divide each section or topic into independent segments that can be added or excluded without diminishing the overall impact. The added longevity and versatility you gain will more than justify the extra time and expense of developing a multi-length, multiaudience production.

Time Schedule

The time between when you begin to plan and the date when the presentation will occur is another aspect of planning that must be considered early in the process. Obviously, this critical interval determines not only your production schedule but your multimedia ambitions as well. The numerous variables for any multimedia production make it difficult, if not impossible, to state specifically how long it will take to produce a project of a given duration. But the more familiar you are with the technical demands, the better you will be at gauging the time.

As a general rule, professional multimedia producers suggest a production-to-presentation ratio of 60:1. That is, about one hour of production time will yield about one minute of a basic multimedia presentation. The ratio climbs as the complexity of the presentation increases. Note: This guideline does not include the time spent brainstorming and organizing your content.

As mentioned, you should allot at least half your available time to the nontechnical aspects of the project, including conceptualizing, planning, and designing. Project management skills are far more critical in multimedia presentations than in most traditional forms of presentation due to the need to acquire and/or create the various elements, design a consistent look and feel, and build in the desired levels of interactivity.

Once you become familiar with the various stages of development and production, you will begin to find shortcuts that are unique to multimedia. The clever use of templates and clip media, for example, can dramatically shorten your development time. Also, once you have completed one presentation, the structure and programming can probably be "recycled," dramatically reducing production time on the next presentation you develop. Spending some time learning about what shortcuts are available before committing to a multimedia presentation is a good idea. This will not only help you avoid a time-consuming reinvention of the wheel, it will spark new ideas and approaches.

tip

Determine the amount of time you will need under optimal conditions to produce your multimedia presentation. Triple it. That's about how long it will take.

WHERE Will It Happen?

The location will determine decisions you make about content, tools, media types, media design elements, and display technology. Basically, you should consider two factors: *audience size* and *environment*. Additionally, you will want to know if the presentation will be delivered on the road, in-house, or both.

For planning purposes, first consider the size of your audience. Table 4-1 outlines the primary considerations both for the display solution and the media design. Use it as a guide throughout the design process to be certain your final product will suit the audience and environment.

Large Audience/Large Room (more than 30 people)	Small Audience/Small Room (less than 30 people)
Less interaction, more formality	More interaction, greater intimacy
Darker room required	Brighter room OK
Large screen size important	Smaller screen OK
Larger fonts on screens	Medium-size fonts OK
More powerful audio speakers	Smaller speakers OK
Overall audience noise a factor	Audience comments heard by all
Lots of small distractions	Fewer distractions
Use more media elements	Less media necessary
Use broad physical gestures	Use facial expressions
Take most questions at end	Deal with questions as they arise
Diverse audience more likely	Homogeneous audience possible

Table 4-1   Presentation Venue Evaluation

During the planning phase, as you are setting objectives, creating a structure, gauging pacing, and selecting media, you should take some time to jot down your delivery/display requirements. You may not have a choice over the location or its characteristics, but by designing the presentation in the context of the environment where it will be delivered, you will be able to optimize the use of media in the production.

HOW Will It Be Produced?

Once you have carefully analyzed your audience, established your objectives, developed your subject, and established your time and place, you are ready to answer the question: *How* will it be done? The "how" part of the process not only involves logistics and production, it is also the point at which you will decide how much outside help you might need.

Multimedia is a desktop technology. Like desktop publishing, one person with the right equipment and content can do it all, from media creation to authoring to delivery. Whether or not you decide to fly solo will depend, of course, on your skills, interest, and time, as well as on the technology available to you. It will also depend on the scale of the presentation.

For simple in-house and basic sales presentations, you could craft a persuasive presentation using only a multimedia-capable computer, an authoring software program, and some clip media on CD-ROM. But for more complex, important, and large-scale projects, you will likely need to involve others. At the opposite end of the production spectrum, you may decide to contract out for the services of a professional multimedia production company. The decisions involved in selecting and working with freelance talent as well as using an outside service are discussed in detail in Chapter 12.

The following sections outline the process of multimedia production for a large-scale project. The rules of good production generally hold true at any level. The procedures and processes can be scaled down to fit any-size production. In addition, if you commit to a large-scale multimedia presentation produced by an outside company, familiarity with the production process will help you know what to expect and will help you get the most for your money.

Logistics and Production

The steps in multimedia production are not significantly different than those for producing a play or a film or any other form of media-oriented entertainment, except that the aforementioned forms are mature and tested; the procedures, protocols, copyright issues, job descriptions, labor rates, and artistic conventions are well established. Multimedia, in contrast, is new and evolving. There is no right way to produce a multimedia presentation any more than there is a right way to make an ice cream sundae. As yet, multimedia has few standard methods, and costs can vary dramatically for services that are remarkably similar.

As you step from task to task toward the final product, keep in mind that none of the advice in this book or any other is immutable. The hardware and software options are changing all the time, as are the aesthetic standards of excellence. Today's whiz-bang, knock-'em-dead multimedia is tomorrow's yawn.

Logistics involves determining and managing your resources: talent, time, and money. This simply refers to the system you will use to make sure that everything is done, on schedule and within budget. Multimedia often involves the coordination of multiple individuals doing multiple tasks. Logistics can become very detailed and complicated. Obviously, the more complex the logistical tasks, the greater the need for thorough planning.

Production involves the actual steps required to complete the presentation. Various models and approaches apply to the production process. Which approach you take will depend on the nature of the project and the available resources. Figure 4-5 provides a schematic of a typical production process that is outlined in the following sections.

Brainstorming

Multimedia planning begins by brainstorming both the concept and content. Draw a line down the center of a piece of paper (or your computer screen), then list every idea relating to content on the left side of the page. If you are developing ideas in a group with others, designate one member as a facilitator to record the ideas. At this point, do not edit any of your ideas. Record everything that comes to mind on the subject, even if it seems inane. Only after you have created a fairly exhaustive list of ideas should you begin to think critically about them. Then, on the right side of the page, rank the ideas according to their utility in meeting

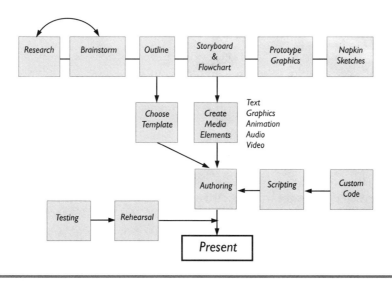

Figure 4-5   The multimedia presentation production process

your goals. How important is it for you to communicate each element of your content? How will it affect your ability to achieve your goals?

Researching

Before or after brainstorming, additional research may be required. This phase of the project should take into account not only the factual research, but also research into what media elements (photos, audio recordings, videos, and so on) are available in support of your presentation. Remember that copyrighted material often carries with it a price for licensing or royalties. So as you research, note the name, address, and phone number of all copyright holders to the works you are considering. This will help you avoid retracing your research path in order to contact the rights holders at a later time (see the "Rights and Royalties" sidebar following).

Be sure to look both inside and outside your company as you research. You may be surprised to find that others inside or outside your department have graphics, photos, written reports, and other materials that will help you. You may even find company videos that you did not know existed. Most very large companies have secured unlimited rights to all kinds of media in the past—songs, commercials, infomercials, and more. The hard part may be finding the right person in the company who is qualified to state definitively what rights are owned. For further discussion of content sources, see Chapters 10 and 12.

Ideas are the milk and honey of effective multimedia. Spend plenty of time brainstorming and researching your subject, and then critically review your ideas. Soon you will have an extensive list of ideas and you should be ready to begin organizing them into an outline.

Organizing

Like any business-related project, presentations need a structure. Whether your concept is narrative (storytelling), expository (informational), or educational, it will need to be organized in some way. Since interactive multimedia allows for branching from one idea to another while presenting, you are not limited to a single organizational scheme. You might use any one or a number of the following:

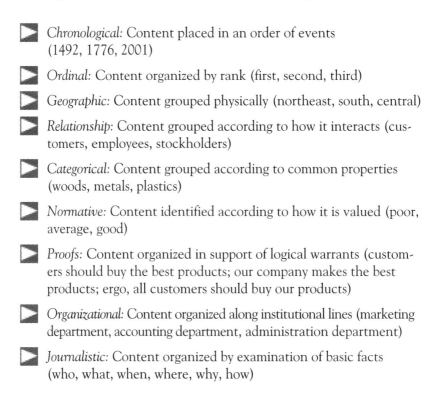

Chronological: Content placed in an order of events (1492, 1776, 2001)

Ordinal: Content organized by rank (first, second, third)

Geographic: Content grouped physically (northeast, south, central)

Relationship: Content grouped according to how it interacts (customers, employees, stockholders)

Categorical: Content grouped according to common properties (woods, metals, plastics)

Normative: Content identified according to how it is valued (poor, average, good)

Proofs: Content organized in support of logical warrants (customers should buy the best products; our company makes the best products; ergo, all customers should buy our products)

Organizational: Content organized along institutional lines (marketing department, accounting department, administration department)

Journalistic: Content organized by examination of basic facts (who, what, when, where, why, how)

Rights and Royalties

The potential for liability in the use of another's work leads to these two simple maxims:

If you did not create it in its entirety or do not have written permission from everyone involved in creating it, then do not use it!

Always consult your attorney.

You have all kinds of rights and licenses to worry about, everything from *model releases* (signed documents giving you the right to use a person's likeness) to *needle drop fees* (payment on a per-play basis). In United States, where freedom of the press and the right to free speech are birth rights, the law draws a hard line at the unlicensed use of intellectual material.

All creators of an original work and their heirs, successors, or assignees retain an interest in the work unless they actively forfeit or assign that interest to another. While the law is sometimes vague and is always evolving, one thing is clear: The damages paid to parties injured when their work is used without permission and/or compensation are often very large. No piece of artwork, music, or video is worth the risk.

Here are a few of the people from whom you may need to obtain rights in a typical multimedia production:

Actors	Musicians
Artists	Photographers
Composers	Producers
Directors	Programmers
Editors	Voice Actors
Models	Writers

Often the collective rights in a work are owned by one individual or organization. Sometimes you cannot know all of the parties who may have been involved in the creation of an intellectual property, so representations must be made by these rights holders that no other parties have any valid claims to ownership of the work. Remember that buying a "copy" of a work or even the "original" medium it was created on does not give you the right to reproduce it or alter it. Only the rights holder may do that.

Rights may be granted or sold for all time, specific periods of time, or a specified number of instances. They may cover all uses of the work or be limited to only very specific uses, such as home entertainment or personal use. Whenever a copy of a work is sold, the law states that it has a "fair use." Any commercial use, such as a business presentation, is normally regarded as going beyond fair use and therefore permission must be obtained and fees or royalties may be owed.

Outlining

By applying organizational schemes to the good ideas on the right side of your brainstorming sheet, you should find that drafting an outline for your presentation is relatively easy.

A useful approach is to outline your material in a manner that looks at the motivational power of the ideas, as well as the logical relationships. For example, you may want to list the most important and obvious selling point of a product first: Top Quality Product. This could be followed by two points of lesser motivational power: Flexibility and Warranty. Finally, you may want to include a kicker that drives immediate action, such as a rebate or a sweetener that increases the perceived value: Early Order Discount.

Give the outline serious editorial review. You should try to receive as much input and put as much editing into your outline as possible. If entire sections of your outline are deemed superfluous, you will save yourself a great deal of effort in the steps to come. Remember, because you are dealing with multimedia, your outline can and should identify known media elements.

tip

Take advantage of the "outline" feature available in most presentation software. Programs such as PowerPoint, Persuasion, Freelance, and Astound use outlines for organizing and generating presentation screens.

A multimedia presentation outline for the earlier pollution control example (in the "Goals and Objectives" section) would look something like Figure 4-6.

Storyboarding

Many presenters view the storyboard as their central planning document. It can be used by writers, artists, and programmers alike as a blueprint for production. Sufficient time should be spent developing the storyboard, because it will guide the creation process and, in some cases, may be the only visual reference other members of the team are given.

A storyboard consists of a series of individual frames, either hand-drawn or created on a computer. Although what comprises a frame may vary, each is based on the smallest full-frame division of your content. If your content is like a slideshow, then each "slide" is shown in the storyboard with its own frame. Video sequences are frequently story-boarded shot by shot for production purposes. In a presentation story-board, if the video already exists, the video sequence would be represented as a single frame. When the frames of a storyboard are

ENVIRONMENTAL POLICY CHANGES
Presentation Outline

I. Opening sequence
    A. Shots of polluted streams and strip mining (*Music*)
    B. Main topic: "Saving the Environment/Saving Money" (*3-D fly-in*)
II. Main menu
    A. Where we are now (*Title slide*)
        1. Waste output
            a) Effluents volume (*Animated chart*)
            b) Regulatory allowances (*Bulleted text with transitions*)
        2. Environmental impact
            a) Known pollution factors (*Video of local conditions*)
            b) Unknown factors (*Bulleted text with transitions*)
        3. Financial impact
    B. Where we are headed (*Title slide*)
        1. New systems
        2. ....

Figure 4-6   Completion of an outline represents the first
significant milestone in most multimedia productions

connected by flow lines and interactivity symbols, the storyboard becomes a flowchart (see the next section).

A thorough multimedia presentation storyboard includes some or all
of the following elements for each frame:

▶ *Key graphic:* The graphic that shows the layout of the frame

▶ *Frame number:* A unique number associated with the frame

▶ *Frame name:* A unique name for the frame

▶ *Media lists:* A listing of each media element to be used in the
frame and its corresponding filename

▶ *Interactivity:* A description of the interactive options to be available for the frame

▶ *Scripts:* Voice-over scripts and video scripts to be used for the frame may be shown or referenced, depending on length

▶ *Speaker notes:* A script or outline of what the speaker will say as the frame appears

Flowcharting

If you plan to include interactivity in your multimedia presentation, you may want to first create a flowchart of ideas. Flowcharting is usually associated with computer programming, but interactive multimedia producers have adapted the planning technique to multimedia. Flowcharts depict visually how the various media elements will be sequenced. The flowchart details the programming logic and the structure of the application.

Unlike the storyboard, which contains sketches of the multimedia elements, the flowchart typically contains only words and symbols. For example, a series of screens that appear in sequence with no option for user interactivity, such as in a linear slideshow, would appear as rectangular display boxes in the flowchart:

Professional programmers use symbols to indicate specific operations or actions when designing a project. Many of the same symbols can be used to create a flowchart for an interactive presentation. Some of the most useful symbols include

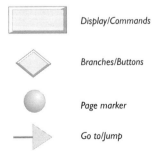

Display/Commands

Branches/Buttons

Page marker

Go to/Jump

Some authoring programs are designed with icon-oriented design structures based on flowcharting. Examples include Macromedia Authorware, Aimtech IconAuthor, and Allen Communications Quest. But buying one of these packages just for the purpose of flowcharting would be extreme. A hybrid form of flowchart includes both visual representations of the presentation's screens and symbols indicating links, jumps, and programming operations (see Figure 4-7).

Unfortunately, flowcharting has no consistent standard. Most professional producers develop their own technique using bits and pieces of different symbols and programs. If the technique you use for communicating the flow of your project is clear for all to follow and uses consistent standards of reference, then it is a good one.

Prototyping

A prototype, or mockup, of your presentation is similar to the rough draft of a document. In it, you will design and test your key graphics and backgrounds, as well as try out sound, video, and interactive elements.

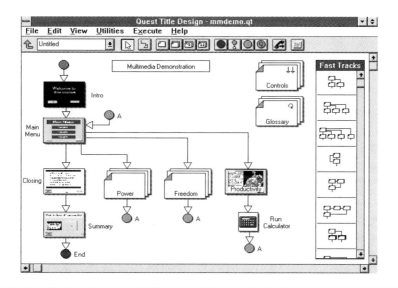

Figure 4-7   Flowcharts depict schematically the structure of an interactive presentation, as in this screen from Allen Communications' Quest

Backgrounds and Key Graphics

Many multimedia programs include templates that can be used as backgrounds for your scenes, slides, or frames. The advantage to using a template included with your software program is that the developers have carefully selected colors and screen layouts, but sometimes you will want to create backgrounds with more texture or dimension than those available in the slideshow programs.

You will need to choose templates or backgrounds early in the design phase of your project and be sure that all elements of your graphics are compatible with them. Any major graphical elements that will recur frequently throughout the presentation should be designed during the prototype phase as well.

Interactivity and Key Elements

Prototyping should also be done to test your interactivity, video, and sound. By testing video and sound file playback, you can check for compatibility or performance issues before exerting a great deal of effort in creating these files. Because system performance and video compression options vary, testing the playback of these files from within your multimedia authoring or presentation software is especially important.

Be forewarned that limitations of your system configuration, especially lack of RAM, will impact negatively on video playback performance (see Chapter 10). So, while prototyping helps to find problems early, it is no guarantee that performance will not deteriorate as file sizes increase.

When testing your interactive options, make these determinations: Is it clear and easy to use? Do pauses occur at the times when the presenter is most likely going to need to expound on a particular point? Does the cursor being used distract the audience or aid in focusing attention?

Scriptwriting

Voice-over scripts, video scripts, and presenter's scripts all need to be written. These may be developed inside your multimedia authoring or presentation software or in a word processor. A voice-over script should include not only the words for the voice actor to speak, but also any special instructions relative to pauses, inflections, or character voices to be used. The video script should follow standard scriptwriting form. Each scene or set should be described. Each shot is scripted separately with designations as to whether it is an interior or exterior shot. Camera instructions, such as pans, zooms, close-ups, and so on, should be included.

When actors are used, the character name should precede the script for that character, and any acting instructions or motivation cues should be included in parentheses.

Collecting and Creating Media Elements

The process of gathering and creating media elements can be one of the most time consuming aspects of production. Not only do you have video production, photo shoots, and sound studio sessions to consider, you can also spend a great deal of effort finding preexisting media for use in your multimedia presentation (as mentioned earlier in the "Researching" section).

Video production and postproduction for multimedia presentations should be done as professionally as possible. Videography is an art form that should not be underestimated by the beginner (see Chapter 10). Whenever possible, hire a professional video producer to shoot your video sequences, and use a professional editor to sequence your shots into usable passages for multimedia. Be sure your video production team understands the overall multimedia framework into which your various video passages are being incorporated. The producer and director should be aware of both the script and the storyboard for your complete presentation.

Voice-over recordings should be "directed" by someone with a critical ear for proper pronunciation and inflection. Even professional voice actors make frequent mistakes, so plan on multiple takes before you get it right. Choose voice actors who can give your presentation the right sound. Your content and concept will help you to choose the voices you need (see Chapter 7).

Capturing Media Elements

The next phase of the multimedia production process is to capture your media elements from their form into digital files that you can manipulate on the hard drive of your system. Photographs are scanned using a scanner; sounds are captured using an audio card; videos are captured using a video board (see Chapter 3).

In order to scan and capture, you must commit to standards for file formats. In the Windows environment, you may choose to create BMP files for your graphics, WAV files for audio, and AVI files for your video. On the Macintosh you may choose to create PICT files for your graphics, AIFF files for your audio, and MooV files for your QuickTime videos. Be sure the authoring system you plan to use supports the import of your chosen file formats. File formats are discussed in more detail in the relevant chapters on graphics, sound, and video (Chapters 8 through 10).

You may need to perform "compression" operations on your media files in order to optimize performance and reduce storage; this will certainly be the case with video files (see Chapter 10). A key component of your planning process will be to allow enough time for video compression–some video compression schemes can take as much as 200 minutes of computer time to compress every minute of video for playback.

It is generally not a good idea to use disk compression with multimedia projects. Most disk compression schemes are incompatible with "on the fly" or real-time playback of multimedia files from the hard disk.

Editing Media Elements

The chapters on graphics, video, and sound cover the details of editing your media elements. What you need to know during the planning phase is how much time editing these elements will take. While generalizing is difficult, you should allow about ten minutes to one hour for editing each scanned image. Audio editing and mixing takes about 3 to 25 times the length of the final recording, depending on the amount of special effects, number of channels, and number of edits. Video editing may take from 5 to 50 times the length of the finished piece. Factors affecting the amount of time required are the number of layers, quantity of edits, and amount of special effects. As mentioned in the previous section, video will need to be compressed, and you should allow plenty of computer time for this procedure, though an operator need not be present at all times.

Authoring

Authoring is the process of sequencing your presentation, laying out media elements on the screen, and entering the instructions for the interactivity you require. With most multimedia software, this process has become quite straightforward (see Chapter 5).

Authoring is the heart of multimedia production, where all of the media elements come together and are assigned a place and time. Be sure to allow plenty of time for authoring, especially if you are new to the authoring software program you are using.

Testing and Editing

To test your multimedia presentation, simply sequence through the frames and look for problems, such as misaligned graphics, wrong files, misspellings and typos, and so on. When you discover a problem, make a note of it and add it to your "to do" list. Testing your presentation often, rather than waiting until the end to discover problems, is recommended. As well, you should always include a testing phase at the end of your project.

Final testing will allow you to determine problems and will be followed by periods of editing where you make changes in your media or authoring. As each editing session is concluded, you will need to test again to see if the problems were corrected properly or if any new ones appeared. Eventually, you must be able to run through your entire presentation without a hitch.

Delivery Environment

Remember, all the hard work that goes into the production phase can be ruined by failure to account for even one of the following delivery factors.

Avoid Direct Light

Sunlight will wash out your visuals, destroying color and readability. If your room has windows, see that they can be adequately draped. Avoid use of lights that radiate directly on the screen. Use a spotlight with barn doors to illuminate the podium, but be sure none of the light interferes with your screen. The ambient light in the room should be limited so that the room is at least semidarkened. Leave enough light for note-taking, if your projection system is strong enough to allow it.

Avoid Sound Interference

Be sure to select a room with no loud, competing noises such as machinery, plumbing, public-address announcements, or telephones, which could diminish the effect of your audio. Beware of locations next to highways or those with doors that open onto central corridors or halls that might be crowded with people at the time of the presentation. Determine what activities, if any, will be occurring in adjoining rooms. Are the walls soundproof? Will your neighbors be using loudspeakers or live music?

Avoid Interruptions

Have someone monitor the door or post a sign. Make it clear to outsiders that a presentation is in progress inside. Likely, you have designed your program to start with maximum impact. Do not start until people are seated. If you are the presenter, have someone else quiet the audience. You may want to run a short still-photo montage or music video to get people's attention. Many presentation software packages have clip sounds that include fanfares or drum rolls, which may seem a bit melodramatic, but it works.

Adjust Room Temperature

As a presenter, you must always remember that if the audience is uncomfortable, your message is less likely to be heard. The room will tend to warm up when filled with people. Be sure your room is thermostatically controlled and that someone is assigned to adjust the heating and cooling if necessary. Keep in mind that a small room that serves well for whiteboard or flipchart presentations can quickly be overheated by the thermal output from projection devices, computers, and peripherals.

Arrange Proper Seating

All the impressive multimedia pizzazz in the world means nothing if your audience cannot see your presentation. Be certain that chairs are set in such a way that everyone can see. Be sure that seats are not placed behind the projection system, where the view may be entirely or partially obstructed. Also check to make sure that the screen is large enough and is set high enough for those in the back to see it. When booking hotel or conference center rooms for a multimedia presentation, be sure to ask the ceiling height and inquire about columns or other obstructions in the room. Inquire about the seating capacity based on unobstructed sight-lines—the fire department–rated capacity is of little relevance.

Have Overflow Seating Available

As with most event planning, you may want to have extra chairs available for any unexpected attendees. Empty seats create a bad impression. Last minute addition of seats, if handled properly, can add to the sense of importance of your message. Have a crew available to add chairs when the situation warrants.

warning

Do not at any time take for granted that someone has already thought of these factors in advance and taken care of them. Whether you are the presenter or the producer of the presentation, you should take personal responsibility for checking the details.

Have a Backup

Accept the fact from the start that multimedia involves a level of complexity well beyond most methods of business presentation. Whether it is your first attempt or your fiftieth production, do not be lulled into a sense of false confidence by the fact that everything is running perfectly before the event. This may sound extreme, but you should plan to have backup copies of all files, including the presentation, applications software, and operating system software. Have on hand, if possible, a backup computer system.

Failing that, be sure to have an alternate form of delivery, such as 35mm slides or overheads and the necessary projector. Some presentation software applications allow for the creation of slides or overheads. They make a useful backup system if something should go wrong with the electronics. With 35mm slides, you can use a device called a film recorder to expose a computer image on a standard roll of film. Certain color printers support the use of special transparency paper for overheads as well. If you do not own a color printer or 35mm film recorder, all major cities have service bureaus and film laboratories that offer these services.

Also, ensure that you have a backup projector or at least an extra bulb for the projection system and someone who knows how to make the change. Have spare computer cables, your own outlet strip with surge protection, and a spare mouse or other control device.

Check Computer Configurations

With multimedia, the safest way is to always use your own computer. Using a computer with a different configuration for delivery will almost always affect the audio and vidio sync. If you have an unpredictable delivery environment, you must plan to author your presentation with a program that can overcome these timing issues.

When you must run your presentation on an unfamiliar machine, be certain to allow time for testing. Check the following items well before the presentation:

- Type and speed of CPU
- RAM quantity
- Free hard disk space
- Type of monitor adapter card (VGA, SVGA, 24-bit for Macintosh 13-inch, and so on)
- Monitor connector type (9-pin, 15-pin VGA, DB-15 Macintosh, and so on)

Check the Sound System

Your multimedia speakers might sound great in your office but fall short when giving your presentation in a different environment. Perform a sound check in the actual presentation space at least an hour ahead of time. Consult with the professional audiovisual staff, if possible. Check microphone mixing and audio levels. Also, try to run through the audio elements of your presentation—sound levels may vary.

Rehearsal

Your rehearsal is different from the testing process. In this final phase, you practice with the most important piece of equipment you have: you! (See the section, "Human Factors" in Chapter 2.) Your voice, your expressions, your timing, your gestures, and your words are still the single most important element of your multimedia presentation. Rehearsals should focus on both content and transitions.

While you may not be able to control all aspects of the multimedia presentation environment, proper planning and production will ensure that you are better able to meet the goals that you have established. Remember, the design of your multimedia presentation is a reflection not only of your content, but also your personal credibility and style.

Production Values

Before beginning any phase of the production process, you will need to determine the overall quality level you hope to achieve. The combination of talent, time, and money you put into a project will determine your final *production values*. Production values is a term borrowed from the entertainment industry. It refers to the perceived quality and professionalism of a television show, play, or film, but it applies equally to multimedia presentation.

The production values you achieve will be a direct reflection of your own standards and those of your organization as well as of how much you invest in talent time and money. In addition to making certain that the production values you set are in keeping with the standards and expectations of your company, you should also take care that you are meeting the expectations of your intended audience. Production values that fall short will derail your presentation's impact. Production values that shoot too far above others' expectations could be a waste of time and money—and a poor reflection on your judgment.

An extremely important, up-the-line presentation to the board of directors or a high-profile annual dealers meeting might justify months of planning, thousands of dollars, and the best talent available. Whereas a simple down-the-line sales briefing for in-house personnel might require no more than a few hours at your desk with a presentation software package. No single formula applies for establishing your production values. You must be the judge of the resources that go into the given presentation. Table 4-2 gives you a general idea of the tools, techniques, costs, and time requirements for producing at different levels of quality. Use it as a guide when envisioning the final product.

Budgeting

As with any media production, your two main budgetary concerns will be the cost of the labor and the cost of the equipment. These two factors, of course, will be dependent on the size and scope of your presentation. As mentioned earlier, you may need nothing more than your own talents and the right technology to create a multimedia presentation, or you may need an outside producer and an entire team of talent. In the sections that follow, you will find references to virtually all the steps and individuals that can be involved in a multimedia production.

	BASIC	**MODERATE**	**HIGH**
Attitude	"Just finish it" "Good enough" "Nothing fancy" "We need it yesterday"	"As good as we can" "Don't go overboard" "Make it work" "Get most for the money"	"No compromises" "Do or die" "Make or break " "The best at any price"
Environment	Meeting room, training room, desktop	Conference room, training room, desktop	Boardroom, auditorium, kiosk, trade show
Time	Days	Weeks	Months
Budget	$100s	$1,000s	$10,000s
Graphics	Simple illustration, template design, clip art	Some custom work; illustration, photos, 2-D animations	Custom illustration, color photography, 3-D rendered animations
Color Depth	8-bit, 256 colors	16-bit, 65,000 colors	24-bit, 16.7 million colors
Audio	Basic quality; 8-bit, 11kHz, mono	Moderate quality; 8-bit, 22kHz, mono	CD-quality; 16-bit, 44KHz, stereo
Music	Effects only	Clip music; clip MIDI	Original composition
Video	Consumer quality; 160×120, 10 fps	Professional quality; 320×240, 15 fps	Broadcast quality; 640×480, 30 fps
Talent	Do-it-yourself, in-house resources	In-house resources, some freelance	Professional artists, designers, programmers
Interactivity	Basic jumps and links	Linking, jumping, branching, hypertext	Fully interactive, custom-programming

Table 4-2   Production Value Guidelines

Note: The rates quoted for labor and wages are ballpark figures; actual wages can vary dramatically from person to person, company to company, city to city, and project to project.

As mentioned earlier, you should first look to in-house sources for help, though the costs for in-house talent must be assessed as well. To get some idea of what components you will have to pay for, refer to Table 4-3. It covers just about all the steps, media types, and talent required to produce a presentation. You can insert these elements as headings in a spreadsheet program, and by assigning costs you will have a basis for your multimedia budget.

JOB CATEGORY	APPRENTICE	TYPICAL	TOP TALENT*
Producer	$10–20/hr.	$30/hr.	$75/hr.
Project Manager	$15–20/hr.	$30/hr.	$45/hr.
Director	$15–20/hr.	$25/hr.	$80/hr.
Writer	$15–20/hr.	$25/hr.	$.50–1.00/word
Storyboard artist	$10–20/hr.	$25/hr.	$30/hr.
Text/data entry	$5/hr.	$10/hr.	$20/hr.
Photographer	$5/hr.	$20/hr.	$50/hr.
Computer artist	$10–15/hr.	$20/hr.	$60/image
Animator	$12.50/hr.	$20/hr.	$500/model
Voice actor	$15/hr	$30/hr	$40/hr.
Actor/Model	$100/day	$300/day	Highly variable
Composer	$15/hr	$25/hr.	$75/finished min.
Audio tech	$7.50/hr.	$15/hr.	$75/finished min.
Videographer	$7.50/hr.	$15/hr.	$250/day
Editor	$7.50/hr.	$17.50/hr	$25/hr.
File conversion	$5/hr.	$12.50/hr.	$25/hr.
Authoring	$12.50/hr.	$20/hr.	$35/hr.
Programming	$15/hr.	$30/hr.	$45/hr.
Audiovisual tech	$5/hr.	$12.50/hr.	$30/hr.
*Rates can go much higher			

Table 4-3   Multimedia Production Wage Rates (average)

Case Study:

Company: Northeast Utilities, Berlin, CT

Business: Full-service public utility company

Objective: Create an interactive employee information system to deliver presentations to employees via a wide area network and promote interactive multimedia's potential for a variety of company-wide applications.

Like many corporations, Northeast Utilities is reengineering its workforce to become more competi-tive. In essence, that means looking for ways to be more productive, but with less staff, and doing so without disrupting employee morale or ne-glecting its customers. In 1993, Jean Lapiene, manager of client services at NU began investigating the use of interactive multimedia as a means of improving intracompany communi-cations and helping the company achieve its productivity goals.

After months of research into the communcations capabilities of multi-media, it was determined that multi-media could help employees who work at geographically separate sites in the New England area to be better in-formed and feel directly connected to each other and the organization.

A team of NU employees work-ing with an outside service provider, IBM in Hartford, developed a com-pany-wide, interactive information network playfully code-named EIN-STEIN for Employee Information News Service Terminal and Em-ployee Interactive Network. While it might seem that coming up with the acronym for the system was the most challenging aspect of the pro-ject, the NU team faced the equally daunting task of setting up seven employee-accessible kiosks at the various utility sites in the Con-necticut and New Hampshire re-gion. Linked by a wide-area network, the kiosks are regularly updated with presentations gener-ated by the NU staff at its company headquarters in Berlin, Connecticut.

The team chose the IBM OS/2 operating system as the basis for the kiosk network primarily because at the time the company's existing net-work was not compatible with Mi-crosoft DOS and Windows. IBM trained two NU communications specialists, Lynn Shea and Linda Classon, in the use of OS/2 authoring tools to enable them to build the presentations themselves and update them as needed.

Northeast Utilities

"We had been doing graphics and newsletters on the Macintosh prior to this," says Shea. "We didn't have authoring or programming experience. We had done a lot of writing. Most of our background was in making content presentable.

"With the relatively steep learning curve of the OS/2 environment, it took Classon and Shea more than two months to begin developing their own content. They authored the presentations on IBM PS/2 486 computers using an authoring tool called MediaScript, and used a combination of Apple Macintosh, Windows, and additional OS/2 software to create the media elements they needed.

The core of their interactive presentation is based on a game board metaphor. Each square of the game board represents a different topic: e.g., "NU in the News," video clips and stories from local news broadcasts; "Online," the company's video newsletter; "401(k) Plan," explaining program rules and benefits; and "Environment," an exploration of the ways NU is working to improve the environment. Einstein, a character that Classon created by making a puppet and then videotaping him in different actions, pops up throughout the program to prompt users on what to do next.

Because the presentations are heavily video-based, Shea and Classon worked closely with the company's corporate video department. They used both preshot video segments and segments created specifically for the interactive presentations.

The video was digitized using Media-Script's video capture tool, then sent to Shea and Classon's department over a local area network. Editing was done using a Windows-based program called Splice. Splice was also used to edit the audio, which includes music, narration (both Einstein and a narrator) and sound effects.

"This required a real blend of skills," says Shea. "There are so many things you have to be able to juggle. You have to know the technology and also know how to communicate, how to present and how to develop content. But I think in many ways the key is the presentation skills. The technology is going to become easier, and people like me who are presentation professionals will be able to do this without being technical wizards."

Outside Services

The cost of services performed outside your company will generally be two to three times higher than the cost to do the job internally. Still, this "out sourcing" is often desirable, because of the considerable time and effort that it takes to gear up a multimedia team. Especially with smaller presentations that employ medium production values, using outside services makes sense.

An experienced multimedia production company is more likely to be aware of techniques and capabilities that can make your message stand out. Nevertheless, the larger your organization, the more likely you will need multimedia presentation services available internally.

You can save a lot of money by bringing as much existing material to your multimedia producers as possible. Not only should you have a complete outline, you should have producers bid the project based on a storyboard. If you cannot create one yourself, pay the multimedia production company to create a storyboard to be used as the basis for bids. For more on working with outside services, see Chapter 12.

The Price of Technology

The cost of technology—the hardware, software, and supplies you need to create multimedia—is a moving target. Your production values will, for the most part, determine what equipment you buy. But trying to determine just what that equipment will cost when amortized over a number of productions is a challenge. Increasingly, organizations have come to view computer equipment as an operating expense rather than a capital investment. In some businesses, the effective life span of computer and other multimedia equipment is measured in months—not because it wears out so quickly, but because something better comes along at similar or better prices.

With this in mind, many financial departments insist that computer equipment be fully depreciated in 24 months. Organizations with high production values should expect the life span of equipment to be about 12 to 18 months. Those with medium production values should plan on 18 to 24 months. In no case should a computer-based multimedia system be depreciated over more than four years. Depending on how your organizational budgets are prepared, you may classify the cost of technology as an equipment expense, capital expense, or as depreciation. In any case, do not forget to account for the short life span of your development and delivery systems in your budget for multimedia.

Budgeting for multimedia equipment is never an easy task. Price changes on multimedia products are announced daily and rumored hourly. It seems there is never a good time to buy, because capabilities are constantly growing with the introduction of new products as the prices of existing products fall.

When budgeting for multimedia equipment, be sure to consider all your options, including using existing equipment, renting, leasing, and purchasing. No single approach will be right for all situations. Remember, the more multimedia presentations you produce, the more it makes sense to acquire your own equipment. Still, effective use of outside services should always be weighed as an alternative to using your own labor and buying your own equipment.

Scheduling

As with budgets, there are no fixed rules for estimating the time a multimedia presentation will take to create. The only rule that can be stated for certain is that it will take you considerably longer than you expect. The scheduling details are usually handled by the multimedia production manager. Sometimes you will be working against a set finish date, other times you will have an open-ended schedule, in which you must insert arbitrary time goals. Remember that in multimedia production, every element is dependent on every other. When one is late, the entire process is delayed.

As in all project management situations, some activities may take place concurrently, while others must follow a chronology. For example, any video recording will have to be done before you schedule the digital capture and compression. Scanning of still photos and digitizing of audio are examples of activities that can occur concurrently, as the completion of one task does not depend on or facilitate the completion of the other.

You must understand these dependencies between tasks, in order to identify "the critical path." The critical path is comprised of any task, which if extended in length will delay the delivery of an entire project. If you are working alone, you can only do one thing at a time, so virtually all tasks are a part of the critical path. But when work is divided among several team members, one team member's work can possibly be delayed without affecting the ability of another team member to begin work.

In order to schedule your production, you should start backward from the delivery date and account for each of these dependencies. Ask yourself what is the step before the current task. And then ask what is the step before that. As you allocate these tasks across time, you may find a need to apply additional resources to certain critical tasks, to keep the multimedia production for your presentation from taking longer than the time that is available.

When scheduling, also remember that few people or organizations actually work at peak efficiency. You must account for interruptions and other mandatory priorities in estimating the number of days that a task will take to complete. Just because a task requires 16 hours of work does not mean that it will be completed in two working days. Meetings take place; the desk has to be cleared; phone calls have to be returned; clarifications have to be made. While good management will try to minimize these factors, they are nevertheless a part of the unavoidable realities of any work environment.

Revisions

Few multimedia producers leave enough time in the schedule for revisions. Revisions, including *debugging* (a programmer's word for correcting mistakes in programming), generating new media, and fixing content irregularities, can be among the hardest phases of production. Some revisions may even take longer than the original work. The best way to avoid numerous revisions is to effectively communicate and anticipate the possible problems. Mistakes are more likely to be avoided if the individual contributors to the project are asked to check their own work. If they are asked to sign off on an approval of their own work, you will be amazed at how careful they are to reduce errors and improve quality the first time.

Revisions occur for three primary reasons:

▶ Failure to thoroughly plan the content of the presentation

▶ Failure to communicate the requirements of the task

▶ Mistakes made by the individual contributors

While good management practices will tend to reduce revisions, revisions are never likely to be eliminated entirely. Therefore, you should insert plenty of time for revisions in your multimedia schedule.

tip

Testing and debugging should be done by the people creating the work as well as by independent reviewers.

Delivery Schedule

The delivery phase of the multimedia project schedule includes the actual delivery of the presentation, set-up time, dry runs, and transportation or shipping time. Allow plenty of time for working through technical glitches. Cables that fail, hard drives that crash and must be restored, unanticipated authoring problems, and other nightmare occurrences must be expected. The more complex the multimedia delivery system, the more you will need to schedule in time for working through last-minute problems.

If you will be presenting in a hotel, for example, your schedule may be cast to the wind by unexpected conflicts with equipment deliveries, union regulations, or accessibility to the room. If your schedule calls for setting up the night before, will you have access when you need it? Will power be available?

Production Strategies

Two approaches are available for the allocation of personnel resources in multimedia production; one is based on the work cell concept and the other is based on the assembly line concept. Thanks to advances in technology and media sourcing, many multimedia presentations are performed by a single individual. The individual is in effect a small work cell.

When a project is being produced by many contributors, the *work cell* is a cluster or team of writers, artists, videographers, programmers, and others. Each cell or team produces part of the presentation. One cell might produce the introduction, while another creates the conclusion, for example. Ideally, the teams work closely with each other to maintain the overall consistency and quality of the production.

In the *assembly line* model, artists work together while videographers work together and yet apart from the artists. Theoretically, the specialization of the assembly line allows for far greater productivity. In practice, however, the enhanced communication of the small multimedia work cells will lead to more cohesive productions that require fewer revisions.

CHAPTER 5

*He who purposes to be an author,
should first be a student.*

—John Dryden

PRESENTATION CREATION: WORKING WITH AUTHORING SOFTWARE

To create a book, you need pen and paper. To create a painting, you need brushes and paint. To create a multimedia presentation, you need authoring software and a computer. Presentation authoring is the pivotal phase of the creation process; authoring software is your pen, brush, hammer, chisel, and welding torch all rolled into one.

What Is Multimedia Authoring?

Think about an author: someone with an urge or a need to create and communicate. Now think about an author at a typewriter. Using the typewriter, the author creates a document, but the information has only one way in, through the keys, and can take only one form, words. Think about an author working with a basic word processor. The information is still limited to letters and words, but the author can import words, sentences, or entire documents in prewritten form. Now picture the author at work on a desktop publishing (DTP) system: integrating words, graphics, illustrations, and photographs into a finished document. And finally, picture the author at work on a multimedia production system: integrating words, graphics, still images, animation, sound, and video into a multimedia presentation.

The tool of the multimedia author's trade is referred to as authoring software. Be aware that some industry pundits insist that a presentation software package is only an "authoring" tool if it has powerful scripting and programming capabilities. But presentation software has matured. The products are no longer just slidemaking and chart-building tools. The multimedia utilities in even the most basic packages provide enough interactivity, media integration, and motion control capability to promote them to the status of authoring tools—or at least junior authoring tools. And that promotes their users to the status of multimedia authors.

Whether you already are or are about to become a multimedia presentation author, you should understand the range of software and functionality available to you. Your software is the environment where you orchestrate all the media elements at your disposal. It is the environment where you both create and construct your presentation. More than any other single hardware or software component, your software program's capabilities (along with your own abilities) will dictate the utility and the quality of your multimedia project.

Simply understanding the features that are available in multimedia presentation software will start your creative juices flowing. You will probably discover techniques and capabilities that you never knew existed, techniques that let you communicate in exactly the way that suits you and your message.

Authoring Functions

Drawing on the parallels to desktop publishing again, recall that DTP software's primary selling points have always been speed, convenience, and cost effectiveness. DTP streamlines labor-intensive page layout, paste-up, and camera work, letting users design, incorporate, assemble, and proof the publication from within the confines of the computer system.

Multimedia presentation authoring tools do much the same, letting you integrate, design, edit, arrange, and synchronize media of different types and from different sources. Much of the speed and creative liberation of DTP programs stems from the freedom to move, cut, copy, and paste elements from inside and outside the document. The same is true for multimedia presentation authoring. Even the simplest programs let you import, export, move, and play media files from different sources. In addition, both DTP and multimedia presentation authoring programs use templates for saving and reusing designs or formats. Both allow for easy updates, alterations, and corrections. And finally, both output directly to the delivery device—printers and presses for DTP, monitors and projectors for multimedia.

Multimedia-capable presentation authoring tools have three basic media-related functions:

▶ *Creation:* Designing and editing text, drawings, charts, graphics, and sound

▶ *Import:* Bringing text, sound, video, stills, animation, and graphics into a presentation

▶ *Integration:* Editing, sequencing, timing, linking, scripting, and playing the presentation

The degree to which you require each of these functions will dictate what application you should use.

Determining the software features you need is largely a function of first determining what you need to do. If you are a chart and graph junkie with little need for interactivity, you might choose a long-standing presentation graphics product such as Software Publishing's Harvard

Graphics with its 160 chart types. Or, if you prefer to develop your graphics and media elements separately and only need a tool for integrating and synchronizing the various pieces, IBM Storyboard Live! or timeline-based MovieWorks from Interactive Solution could do the job. Perhaps you already have a number of presentations created in one of the top-selling presentation packages, such as Microsoft PowerPoint or Aldus Persuasion, and you want to integrate them in a package that gives you superior animation capabilities; your answer could be an application such as Vividus Cinemation.

Authoring Software Packages

Software packages that are capable of multimedia or have been designed specifically as multimedia authoring tools fall into roughly three categories:

▶ *Multimedia-capable presentation software:* Most of these programs were originally designed as charting and graphing packages for slidemaking, but now include various levels of electronic screen-show and multimedia functionality, from simple media linking to complex media integration. They can create some media elements, such as charts, graphs, and simple illustrations, but they generally have no built-in tools for recording audio or video. Examples include Microsoft PowerPoint, Aldus Persuasion, Lotus Freelance, Micrografx Charisma, Software Publishing's Harvard Graphics, DeltaPoint DeltaGraph Pro, Alpha Software Bravo, WordPerfect Presentations, and CA-Cricket Presents. A few paint and drawing programs, such as Corel Corporation's Corel 5.0, also include a charting and screenshow module with the program, putting them within the category as well.

Newcomers to the category, such as Asymetrix Compel and Gold Disk Astound, have maintained charting, graphing, and slidemaking capability, while focusing heavily on the tools needed for interactive multimedia production. Others, exemplified by Macromedia Action!, AskMe Multimedia SST, and Q/Media Software's Q/Media, concentrate on multimedia screenshow capability while maintaining many of the familiar graphics tools, but they are not intended for output to 35mm film or overhead transparencies.

▶ *Dedicated media integration software:* This relatively new breed of presentation software was designed strictly to import, edit, sequence, and play back multiple media types. Most of these programs lack the powerful graphics and output features found in general-purpose presentation programs. They typically do not offer the tools to create media elements, but they are capable of importing a wide variety of sound, video, graphics, and animation file formats. Examples include IBM Storyboard Live!, Lenel Systems MultiMedia Works, Asymetrix MediaBlitz!, and Passport Designs Passport Producer.

▶ *Professional Development software:* These complex and costly tools are used by professional developers for multimedia presentation authoring, interactive consumer titles, and interactive courseware development. None of the products in this category are for the casual user. They are characterized by a built-in programming capability and such sophisticated tools as response judging and data collection. Examples include Macromedia Director and Authorware, Asymetrix Multimedia ToolBook, Aimtech IconAuthor, Apple HyperCard, and Apple Media Kit.

Each level and each software package has strengths and weaknesses in different aspects of multimedia production. Each has different user interface designs to be considered, as well as different operational models. In the end, all have the same goal—the creation of successful presentations—but each shoots at the target with a different arsenal. Indeed, the available authoring tools, costing anywhere from $80 to $8,000, range from presentation peashooters to multimedia Big Berthas.

When making your determination, you should look at not only the features, price, and ease-of-use issues, but also at your own time constraints and production talents. If time is short and you have a low threshold for learning new software, then entry-level (see Figure 5-1) and some midrange products are available that can help you get a multimedia project up and running in the shortest possible time; but you should not expect MTV-quality results.

The following overview of presentation authoring tools will give you a good feel for the different types and tiers of products and the skills required to use them effectively. But only careful consideration of each individual program and hands-on experience will tell you if you have found the glove that fits your hand.

note

Several of the most popular and useful presentation authoring packages are included on the CD-ROM. Because selecting one of these programs is as much a question of personal style as of the intended application, the CD-ROM gives you an ideal environment to test-drive the different products before you buy.

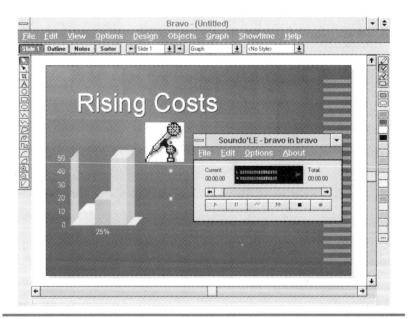

Figure 5-1   Even low-cost, entry-level presentation packages, such as Alpha Software's Bravo ($79), feature digital video import and sound capability using object linking and embedding

Multimedia-Capable Presentation Software

Unless you have been living in some remote cave, beyond the reach of power lines, you have experienced, and probably delivered, a 35mm slideshow. Slideshows, with their familiar *cachitika-chuk, cachitika-chuk* as one slide is ejected and the next slide drops into place, are one of the most popular forms of presentation ever invented. Presenters everywhere understand how to mount slideshows: generate the slides, number them, key them to your script, slip them into the carousel, and pray the bulb doesn't blow.

Considering the success of the format, it is no surprise the first software tools for designing and delivering presentations from the computer adopted the slideshow as their metaphor. As noted, these programs were designed originally for creating graphics and outputting them to a film recorder to generate 35mm slides—still their predominant use today.

Slideshows are primarily *linear* events. That is, they run in a straight line from beginning to end (with the exception of a few reversals or an embarrassing pause while someone turns a slide right side up). 35mm slideshows are also *static*. Slides do not walk, talk, sing, or tap dance. Each image is complete and self-contained the moment it first appears. Regardless of how long the image stays on screen, in order to change the information, you must change the slide.

Once you move from 35mm slides to a multimedia screenshow, you are no longer slave to the predetermined order in the carousel or at the mercy of the slide service bureau. In practice, the slideshow may not be the best model for multimedia presentation, but it has an inherent comfort factor. It provides a logical and nonthreatening transition from familiar presentation techniques to electronic delivery.

Many of the mainstream presentation software packages, such as Freelance, Charisma, Persuasion, and PowerPoint, employ the slideshow metaphor. They borrow the concept of the light table for sorting (see Figure 5-2). They also use numbering and annotation schemes that mimic the way a presenter organizes a projector-based slideshow. In addition, they have all the necessary functions to help you output individual images to 35mm slides and overheads.

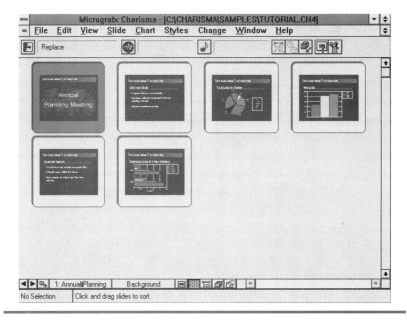

Figure 5-2   Micrografx Charisma uses the slideshow metaphor
literally, adding 35mm slide-like frames around
images in the slide sorter view

As you will see demonstrated in the following pages, dynamic elements such as sound, video, and animation are treated by the software
interface as "objects" to be imported and integrated onto a slide. Multimedia presentation authoring at this level, then, is nothing more than a
slideshow that walks, talks, sings, and tap dances.

Features and Tools

The media creating and editing tools within multimedia presentation
software vary according to what the program is ultimately designed to do.
Most general-purpose packages offer extensive text handling and graphics tools, as well as utilities for automatically generating charts and graphs.
A few go so far as to let you record and edit sound, and at least one
program has controls for capturing and editing video.

Outlining

The single most important feature in multimedia authoring (though it is not technically a feature at all) is that it should think and work the way you do. You should not have to radically restructure the way you develop and record ideas. You should not have to spend weeks trying to learn a new way of converting your written concepts and information into a visual format.

If you regularly organize your ideas in outline format, and you want the quickest route to a series of slides built from your outline, be certain your multimedia presentation package has an outlining feature. Most of the traditional presentation software packages, as well as new multimedia-capable presentation software such as Astound let you either type an outline directly into the presentation or import your outline from a word processing program. The heads and subheads of the outline are automatically inserted into the appropriate positions on each slide using the template or master slide as a guide. Look for a package with an onboard spell checker, particularly if you do most of your writing within the authoring program and not in a separate word processor.

Sorting and Arranging

You have a number of ways to sort and rearrange the order of your slides or scenes as you work, but the most familiar is the thumbnail or slide sorter view. Found in some form on all of the general-purpose presentation graphics packages and all but a few of the media integration packages, slide sorters display each of your images as small sketches. By clicking and dragging, you arrange them into a new order or eliminate them from the presentation sequence. (See Figure 5-3.)

Templates and Master Slides

The rules of good design demand that your presentation have a cohesive appearance. The layout, text style, and color scheme should be consistent from slide to slide, and elements should connect thematically. The best packages let you design a master slide or pick one from a gallery of predesigned templates, then they automatically apply the elements to the rest of the slides or scenes in the production. Look for packages that let you preview template families. *Families* are sets of templates that incorporate the central design in a variety of slide types: opening screens, text screens, video-in-a-window screens, and so on. (See Figure 5-4.)

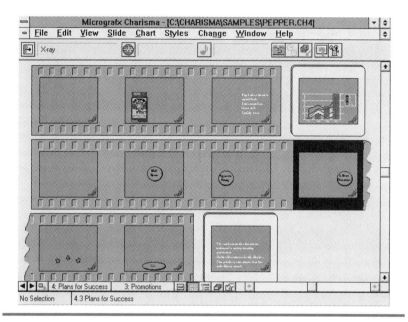

Figure 5-3 Charisma sorts animation sequences by connecting them as individual frames on a roll of film

Painting and Drawing

Integrated paint and draw tools in multimedia presentation software vary from none to near-professional-level utilities. The tools you will need are strictly a matter of preference; but generally, if you do most of your graphics work outside the program, you will not need much capability internally. On the other hand, if you want to avoid the need for outside graphics tools, you should look for a package that will let you paint and draw at whatever level of artistic quality you require.

A few packages, such as Persuasion, have taken the artistic high road by incorporating detailed and highly functional tools suitable for use by a graphics professional. Charisma's latest version includes 3-D perspective and modeling tools commonly found only in professional computer graphics applications. PowerPoint and a few others add an additional "screen draw" feature that lets you annotate images on the screen during a live presentation using a special drawing tool. Be aware, a few

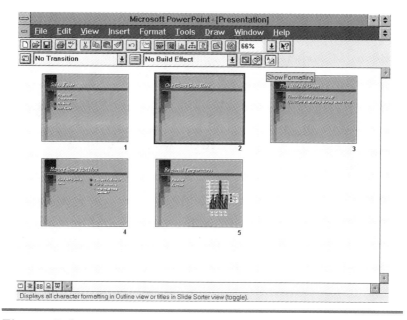

Figure 5-4   PowerPoint lets you pick a design, then offers a variety of layouts for different slide types

of the software packages give you no more than the basic shape drawing and coloring tools to be found on any gradeschooler's computer program, a situation you may come to regret as your drawing and painting skills improve.

A couple of products, notably the suite of tools from Corel Corporation, use a modular approach to graphics tools. The painting, drawing, charting, animating, and presenting modules are sold together but operate as stand-alone products. Painting and drawing tools are discussed at length in Chapter 8.

Charting and Graphing

With powerful charting and graphing tools now included in everything from spreadsheets to word processors, the days of hand-drawn charts and graphs are all but over. Just how much charting power you will need will depend on what type of information you typically convey and how much sophistication you need to pack into your charts.

Harvard Graphics, Stanford Graphics, and DeltaGraph Pro are three packages well known for their charting functions. Freelance, PowerPoint, Charisma, and others also have built-in charting functions to suit the most demanding presentations. Animated charts are becoming the de facto standard in onscreen presentations, so look for a package, such as Astound, that features automatic graph animators.

Object Linking and Embedding (OLE)

All but a few of the programs now available support Object Linking and Embedding (OLE). Pronounced oh-LAY, the standard has, in effect, shifted the nature of presentation authoring from an application-centered view, in which everything happens within one fixed program environment, to an event-centered view, in which the presentation becomes a "living document" that connects with other applications. That is, the information in the presentation changes along with changes that are external to the software.

This breakthrough in conceptual design should not be underestimated. By linking data and media elements that have been created at different times and reside in different applications, presentations are transformed from one-of-a-kind shows that begin going stale the day they are delivered, to fresh, dynamic, and constantly changing communication events. In addition, linking lets you use the original media object in any number of presentations, all of which will be updated as the original is revised.

Using OLE, media objects—graphics, sound, video, animation—can either be embedded or linked to the presentation. *Embedded objects* are actual copies of the object that reside in the presentation authoring program separate from the application in which they were created. They maintain a connection to the original application, but *not* to the original data file. Consequently, any subsequent changes in the original object are not reflected in the presentation unless a new copy is made and embedded.

Linked objects, by contrast, are representations of the original object. The linked object exists outside the authoring application and can be shared by other applications or presentations. Linked objects require the presentation program to keep track of the object file where it lives in the original application. When the original changes, the author has the option of keeping the image in the presentation the same or updating it.

Because a linked object does not "live" in the presentation, the authoring program only knows its "address." When it comes time to find the object, the program will go calling. But if the object file path has been corrupted or the file is missing, nobody will answer the door; the house will not even be there. With embedded media, this is not a problem because the object exists within the presentation. Fortunately, OLE-capable software is getting smarter. Many authoring programs now have built-in media locators that track down fugitive media. (See Figure 5-5.) Some of the more advanced locators will even consolidate a presentation's linked media into one centrally located folder for easy access.

File Formats

Multimedia opens a Pandora's box of file types and options that your software will need to address. Most packages support an assortment of graphics file formats as well as several sound formats and digitized video, typically as QuickTime movies (MOV) or Video for Windows (AVI).

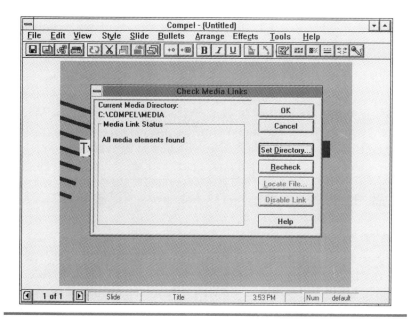

Figure 5-5   Compel checks to make sure all linked files are available to the presentation

While dozens of file formats are in use, the need for interchanging files between different software tools has lead to a certain degree of standardization. These standards are supported by the major tools for creating, authoring, and editing, making the process of integrating media much simpler today than it was even two years ago.

Each media type has at least one standard file format. The standard can be different depending on whether you are using a PC or a Mac, so do not assume that a file created on one platform can be automatically read on the other. The file formats for graphics, sound, animation, and video are discussed in Chapters 8, 9, and 10.

Timing and Sequencing

With dynamic media elements, the order of events within a slide or scene can be as important to the overall effect of the presentation as the order of the slides themselves. In the more traditional presentation packages, the timing of media events is dependent upon the length of the media clip that is being played. Transitions, slide builds, and other basic motion effects occur at set rates. A few packages allow you to set the time for an event, but it can be difficult to coordinate or synchronize the actions of multiple events in one slide or scene.

The newest crop of multimedia presentation programs, such as Action, Q/Media, and Astound, incorporates timelines for synchronizing individual objects. A *timeline* displays the screen object (text, picture, video, sound, and so on) as an item on a horizontal track or bar. Vertical lines divide the scene into units of time. The duration of the object's time on screen as well as its actions can be adjusted in relation to the total duration of the scene. In programs such as Action and Q/Media, the object tracks are visible on the same timeline, making it relatively easy to time their actions in relation to other objects (see Figure 5-6).

Animation

Simple slide transition and animated bullets have been part of electronic slideshow presentations for years. Not until relatively recently, though, as screenshows have grown in popularity, have they been given much attention. The eye-catching motion produced by the changes between screens and the ability to animate bullets and logos has become

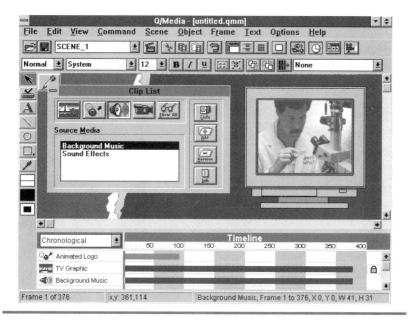

Figure 5-6   Q/Media uses a timeline to synchronize media objects within scenes

a staple for good electronic presentation. A few packages let you assign animation to a chart or graph as you build it. *Screen builds,* the effect of adding elements to successive slides, is now available on most packages as well.

Using import and linking utilities, most packages are capable of playing back animation files created outside the program. Newer multimedia presentation software has gone a step further by allowing you to create animations within the programs themselves. Animation capabilities vary greatly. Action, for example, lets the user create standard slide builds, animate charts automatically, or make objects move on cue along a predefined path. For more sophisticated animation techniques, such as the cel animation used by professionals, packages such as Cinemation allow you to both move an object along a defined path and create objects with individual movement. (See Chapter 8 for a discussion on animation.)

Interactivity

Multimedia presentation by its very nature embodies interactivity. The ability to import or link to other media, even other applications, and to control those links both during the authoring process and during the presentation, is central to the persuasive power of multimedia as a business communication tool.

Despite the fact that the slideshow metaphor implies linearity, delivering electronically and in digital form allows for a certain amount of nonlinear movement or interactivity. The most basic form of interactive control is *pacing*. As with standard slide presentations, the speaker or an assistant advances the slides on cue. This may not seem to fit a strict definition of interactivity, but it does imply that the presenter can react to the audience by speeding up, slowing down, or pausing. All of the presentation programs on the market allow you to advance your presentation manually or set time intervals for automatic delivery.

Another form of interactivity, linking media files, provides not only more dynamism to the presentation but also greater depths of interactive information. When authoring a presentation, you are given the option of having the media information play as the slide is displayed or of manually initiating the media event at some point while the slide is on the screen. You might, for example, choose to play a video, audio, or animation segment only if the audience expresses an interest in the extra information it offers. Media controllers that either come with the computer system software or with the presentation program give you rudimentary control of the media event, usually with forward, reverse, stop, and start controls similar to those on a VCR or tape player.

A few of the presentation packages also let you create hot buttons and assign jumps, making it possible to create basic branching interactivity. Even the veteran presentation software packages, such as Harvard Graphics, Freelance, and WordPerfect Presentations, will let you set up links for jumping and branching between slides using special "hot" buttons or other action objects. The newer multimedia presentation packages go further. Compel, for example, lets you define any word or object on the screen as a hot button. You can assign the objects actions, ranging from playing an audio file, pausing for a user-triggered action, or leaping to some other part of the presentation.

Output Options

One of the central delineators between traditional presentation software with multimedia capabilities and software designed specifically for multimedia authoring is the type of output each can produce. Slides, transparencies, and hard copy (paper) are the bread and butter of traditional presentation packages. The programs are designed to optimize the image creation process so that colors, dimensions, and resolution quality stay intact. Many of the programs provide built-in utilities for preparing slides for electronic transmission to a slide service bureau.

If you know that you will need slides, transparencies, or color printouts of your presentations, you would be wise to stick with one of the many packages that have specific utilities for the process. Trying to save an image of a single frame from Action! or MediaBlitz! to be output as a 35mm slide can be a study in frustration. A better course would be to use a product such as Compel or Astound. Astound in particular provides all the necessary output tools, including options for transferring your multimedia presentation to videotape, provided you have the right hardware installed.

User Skills

General-purpose multimedia presentation authoring programs are designed to serve both the novice and veteran presenter. They each attempt to offer the maximum level of functionality with maximum ease-of-use. Some users report that they have adopted their presentation software as a primary tool for writing memos and short reports as well as for general project management—the material they create is instantly available in a graphic format that can be played on screen or printed to film or paper.

Anyone with a basic understanding of standard office software, such as spreadsheets or word processors, already has most of the skills needed to use the products. In fact, many of the more powerful office productivity packages now contain features that could once only be found in presentation software. Many word processors, for example, can take advantage of OLE to attach media objects. Some spreadsheets also contain OLE functionality and many offer strong charting and graphing tools as well.

When considering a presentation software program, you may want to look at one of the several office productivity suites on the market.

These products bundle presentation software with such applications as spreadsheets, word processing, database, e-mail, and project management. Typically, the products in these groups are designed to work together, easily sharing files and setting up links, using internal tools that operate similarly from program to program. Lotus Freelance, for instance, has designed in direct links to the Lotus Notes database program, allowing entire presentations to be stored, sorted, and retrieved.

Pricing

Street prices for multimedia-capable presentation software range from about $80 to $600. If you are already using a presentation software package, keep an eye out for competitive upgrades. In some cases, the vendors will offer a package at up to two-thirds off list price if you turn in your old software. The price cutting for these products becomes intense at times, so watch for the deals.

Unlike many hardware categories, the higher price does not necessarily mean more or better features. As mentioned, each package varies slightly in its extended features. Choosing one is a matter of knowing who will be using the package, how often, and for what level of presentation.

The two products that most successfully bridge the gap between slideshow software and media integration tools, Compel ($295) and Astound ($399), are priced comparably to slideshow software. Q/Media, which is less powerful but compares favorably to Astound and Compel in some areas, sells for less than $200. Macromedia's dedicated multimedia presentation package, Action!, lists for $299.

tip

Buying a presentation software package as part of a suite greatly reduces the net price.

Media Integration Software

In the past two or three years, the proliferation of digital sound and video products has spawned a new breed of multimedia authoring software. Designed from the ground up for creating multimedia productions, media integration packages have little or no focus on creating images for output to slides or overhead transparencies. They are not designed to create charts or graphs. Most have abandoned the slideshow metaphor in favor of models that conform to the production of the dynamic, time-based media they are designed to handle. Several have borrowed the editing and assembling metaphor from the medium of film and video.

Full-fledged media integration programs do almost nothing but import and organize media elements. MediaBlitz (bundled with Compel) is an example of a program that has no tools at all for creating media but is proficient at putting media together in a timeline environment and synchronizing the playback. The MediaBlitz timeline covers the entire screenshow, or one file. Multiple files can be played in a selected sequence or exported to Compel for incorporation into a presentation. MultiMedia Works from Lenel Systems International is another example of an integration program that can import, sequence, and play back almost any media type but has no graphics tools (see Figure 5-7).

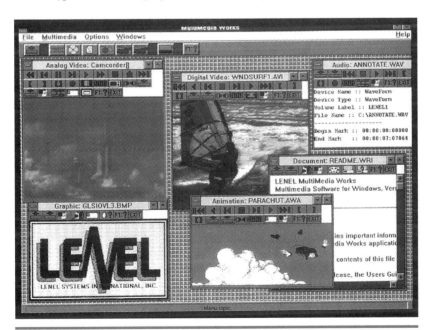

Figure 5-7 Lenel Systems MultiMedia Works integrates and plays media that has been created outside the program

Case Study:

Company: Xircom, Inc., Calabasas, California

Business: Portable networking products manufacturer

Objective: Author a modular digital presentation that can be customized and adapted for different events by the presenters themselves.

In the past, Xircom, Inc. had used traditional presentation materials, such as slides and overheads, for its trade show demos, dealer training, and sales meetings. The company realized the process was wasting company resources and was creating undesirable inconsistencies from one presentation to another, because it caused materials to be created redundantly by the various departments. Digital multimedia technology, it was decided, would let the company develop a single set of materials that could be adapted uniformly and quickly for virtually any presentation situation.

The company designed a modular series of text, graphics, sound, and animation elements that are avail-able to all its sales, marketing, and training personnel.

The first module consists of a complete company background, including the history of its current product line and industry awards.

The second module, intended as the core of the presentation system, discusses each of Xircom's three primary product lines in detail. A third module teaches Xircom salespeople the finer points of selling the company's products, and a fourth module contains the creative elements for a corporate marketing campaign themed around major league football.

The modules are tied together thematically by an animated cartoon character named Adapter Man, who guides the user through each of the presentations and demonstrates procedures, such as how to connect a PC to a local area network using Xircom's pocket LAN adapter. Lacking in-house multimedia production capability, Xircom hired Digital Beach, the creative services division of Santa Monica, California–based HSC Software to author the materials.

Xircom, Inc.

Digital Beach breathed life into Adapter Man, first by hand drawing cels of the character, then scanning them into a 486/50 PC. The images were converted into Autodesk's Animator Pro GIF file format. A 3-D background, created with Autodesk 3D Studio software, was added be-hind the Adapter Man animation, allowing the character to, among other actions, hop on a desk and interact with the actual products.

Digital Beach authored the modules using its own HSC InterActive software. The software uses an object-oriented structure that lets users program operations and links by merely dragging and connecting icons. The specific material for each icon, or object, can be altered without affecting the overall structure of the program and vice versa. To create a presentation, the presenter selects icons from the collection of elements and assembles them in any order and with the desired degree of interactive links. To take advantage of the software's capabilities and give each of the departments in the company maximum versatility when creating presentations for individual events, Xircom employees were trained to use the authoring software themselves.

Costing about $50,000 to produce, the presentation in its various incarnations has made appearances in Europe and throughout the U.S., aiding Xircom executives and sales and marketing staff to deliver their message of ease of use and portability.

At the high end, products such as Passport Producer Pro from Passport Designs can not only integrate media elements using professional video and audio industry time codes, they can control external video and audio sources for playback during a presentation.

User Skills

Skill levels for using media integration tools are not significantly greater than those for traditional presentation software. The main difference is that timeline-based products force the user to view media events in a new, dynamic way. This is not always easy for people who are used to thinking about one image or media event at a time. Some animation sequences, for example, move several objects simultaneously, requiring that you understand the relationship between time and the visual illusion of motion.

If you are intimidated by the timeline metaphor and prefer to stick with the idea of a slide, your best bet will be one of the programs that remains loyal to the slideshow metaphor. If, on the other hand, you are a timeline thinker or a closet filmmaker, you may want to go straight for a media integration tool to give you maximum creative control over your media synchronization. Essentially, it comes down to a matter of style and thought process.

Moving into media integration tools means upgrading your skills in several areas. First, a solid understanding of the different types of file formats is needed in order to successfully import files into the presentation. While you will not need a technical understanding of the bits and bytes of file structure, you will need to be able to recognize the various file types. For example, many integration tools will import a graphics file of the PICT type but will not recognize an EPS file. Knowing the technical terminology is a key to integration tools. A working knowledge of file conversion utilities will be of considerable value as well. Further, you will need to develop skills in media creation and editing using tools external to the media integration program.

Pricing

While media integration tools take a different approach to building and delivering presentations, they do not, in general, cost more than traditional slideshow presentation software. Prices run from about $200 to $700, but the cost of other necessary programs must be factored in as well. A product such as Cinemation offers above-average animation tools, good interactivity features, and good paint and draw tools for $495, but it has no outliner and is not adept at creating such simple items as bullet slides. You would likely need to own a separate presentation software package as well.

Professional Development Software

Professional multimedia development software represents the high end of the authoring tools scale and, as such, comes with several caveats about skill level and time requirements. These are the tools used mostly by professional production companies, CD-ROM title developers, and professional trainers. While prices can be as low as $500 for some products in this class, not even the vendors who sell them pretend that these are products for the casual business presenter.

Central to the functionality of all products in this category is their ability to create sophisticated interactive presentations using embedded programming languages. Most can also incorporate code from external programming languages.

The differences between simple branching capabilities offered by many media integration tools and fully interactive and programming features are significant. First, a fully interactive presentation is often delivered to the intended audience for viewing without a guide, such as a speaker. This means that the absentee presenter must anticipate what the user will most likely want to do at each point and provide a mechanism for the user to execute those wishes.

Second, because the user will be interacting with a computer system in order to, in effect, deliver the presentation to himself, feedback must be provided to the user for guidance and evaluation, a technique know as *response judging*. Additionally, these presentations often seek to gather data about the user for later analysis. This information could be as simple as name and address or as complex as tracking choices to learn more about the user's background and interests. All of these operations require some level of programming.

What follows is a brief overview of the types of professional-level programs that are available.

Card Stack Model

A few professional-level multimedia authoring programs are built around the idea that each screen is a single page in a document or an individual card in a stack of note cards. Each card can hold graphics, text fields, and one or more "hot" buttons that trigger a media event or take the user to another card. When creating a presentation in a card-based model, a card is created for each screen event. Typically a main menu or hub card serves as the focal point for navigation. Apple Computer's HyperCard is the most familiar example of a card-based authoring environment. It should be noted that HyperCard and programs like it are not specifically designed for multimedia presentation. They lack many of the ease-of-use and graphics features that have become standard for presentation authoring. HyperCard's real strength is its programming and scripting capability, a feature usually reserved for professional developers or presenters who have a flair for programming.

Asymetrix Multimedia ToolBook, a popular professional-level authoring tool, opts for the page metaphor. Each "page" in the multimedia document holds a set of objects, media events, and commands. The authoring process involves moving to and from the pages using programmed links and operations.

Cels, Frames, and Channels

The use of cels, frames, and channels as the individual units of organization for multimedia authoring was popularized by Macromedia in its professional multimedia development software, Director. Director's metaphor differs from most in that the screen is treated as a stage on which actors (objects) perform. The performance is controlled by generating a new "frame" every time a media element is altered, added, or removed from the screen, similar to the way a movie scene or an animation is built frame by frame. Frames can be displayed with pauses in between, as in a standard slideshow delivery, or they can be played back at a rate that simulates motion. For purposes of interactivity, the presentation can be programmed with Director's built-in scripting language to move to any frame.

In Director, a presentation is created on a grid-like system called a *score*. A frame is made up of vertically stacked cels, each representing the status of an object on the stage at that point. The horizontal rows, or channels, represent the changing status of the media objects or actors from frame to frame (see Figure 5-8). A primary advantage of frame-based tools is the creation of events and total presentations that are precisely (to the frame) synchronized. The frame-based model works well for animation, because each movement can be adjusted on a cel-by-cel basis. In fact, Director is often used as a stand-alone animation tool.

Icons

A few multimedia authoring tools use an icon- or object-based model. The screen images and events that in other packages are represented as slides, scenes, or sequences are depicted as small pictures or icons. The

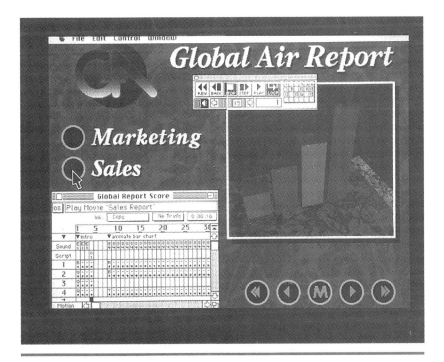

Figure 5-8   Macromedia Director creates frames (columns) that are orchestrated on the spreadsheet-like score

icons provide a visual representation of the structure of the presentation. To build the structure, you drag the icons into a desired order and link them together with lines or arrows, much as you would in building an organizational chart or flowchart. Clicking on an icon opens the image and displays it on screen for playback or editing. This technique is a favorite of multimedia developers who need a high degree of interactivity.

The icon metaphor is used mostly in high-end and costly professional multimedia development tools, such as Macromedia Authorware (see Figure 5-9), Aimtech IconAuthor, and Apple Media Kit. One of the few exceptions, HSC InterActive, is a stripped-down version of IconAuthor. It reduces the price and the feature set of IconAuthor to bring icon-based authoring to a wider user base, but remains a challenging program to learn and operate.

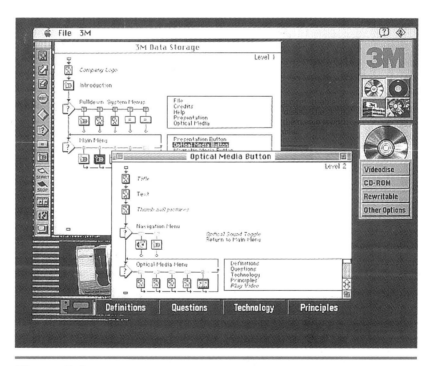

Figure 5-9   Macromedia Authorware for professional users employs an icon-based authoring model

Licensing and Run-Time

For corporate users, run-time license fees are typically calculated as an annual fee. Payment of the fee permits the use and distribution of an unlimited number of run-time projects. Most general-purpose presentation software and media integration packages do not require license fees.

A clear understanding of the licensing fee schedule for your tool of choice is critical. The closer you get to full interactivity, the more likely you will be required to pay a license fee. Be particularly aware if you turn your presentation into a widely distributed or sold product. Commercial uses of these authoring tools are even more likely to require a license fee. When in doubt, check with the

sales or marketing department of the tool vendor.

For purposes of distribution, most presentation authoring programs include *run-time players* as part of the package. Run-time players or run-time engines, as they are sometimes called, are sections of programming from the authoring application that contain enough code to run the presentation. Run-time players are only a fraction of the size of the complete authoring program. They can be distributed as part of the multimedia presentation or they can be distributed separately to the user. The user need not have access to the original authoring program at all in order to play the presentation.

Some media integration tools, and many fully interactive authoring tools, require the payment of a run-time license fee for applications that you distribute. These fees are rarely assessed on presentations that are only given one time, but if you intend to distribute your presentations to a sales force or place them on a kiosk, license fees may be due the vendor.

Pricing

Most high-level authoring programs are expensive. Macromedia Director, one of the most widely used authoring tools, lists for $1,195. Asymetrix Multimedia ToolBook, a market leader for PC-based projects, is $695. Professional products designed more for courseware development are a major investment. Macromedia Authorware and Aimtech IconAuthor each list for $4,995. The new Apple Media Kit is priced at about $1,800. A couple of the more affordable products in this category, though far less powerful, have already been mentioned: HyperCard lists for just $199, SuperCard costs $299, and HSC InterActive sells for $295.

Authoring Tutorials

Multimedia authoring software selection is the most important decision you will make in the multimedia presentation process. Whenever possible you should try before you buy. The software provided on the accompanying CD-ROM is a good place to get started if you are new to multimedia presentation software. It is also an excellent resource in case you are considering an upgrade or a change to another package.

The following two tutorials will introduce you to the two most common models used by multimedia presentation software: slideshow and scene building using a timeline. As with the other tutorials in this book, these are not meant to teach you how to use the software; you will need a manual and some concentrated effort to do that. They are designed to walk you through the first basic steps for creating a presentation using media other than basic graphics or charts.

Both Lotus Freelance and Macromedia Action! are included on the CD-ROM for you to try.

Multimedia Slideshow: Lotus Freelance

Freelance Graphics for Windows was one of the first presentation graphics packages on the market. It was innovative in its day and continues to be a strong and evolving software program. Like most of its kind, Freelance has a wealth of outlining, sorting, charting, and designing features. Its emphasis is on ease of use. The program guides you through the entry of presentation content using text boxes and templates.

Multimedia is not a focus of the program; its central purpose is the creation of slides, overheads, and hard copy. But it is a good example of what general-purpose presentation graphics software can do with multimedia. You will find a central work area, pull-down menus, customizable icons, and a status bar from which many editing functions can be performed.

Freelance, like most presentation graphics packages, lets you embed or link to multimedia objects using commands from the Edit menu. You can also quickly access objects via the icon for multimedia. In most cases, a symbol or icon representing the embedded object or an object link is inserted into the slide. Once an object has been added to the presentation, it can be cut or copied. The symbol or icon representing the object can be moved, aligned, stretched, contracted, or hidden. Changes to the object itself require editing in an application or with a device that can utilize the file format of the object.

In this scenario, you are an operations manager preparing a presentation to be delivered to upper management on the subject of team building. You want to go beyond a silent, static presentation, so you will be adding the multimedia element of sound.

Figure 5-10   Create a bullet slide using one of the available SmartMaster templates

Step 1: Launch Media Manager

Start Freelance and create a new presentation with one of the SmartMaster templates. The one chosen for this example is finance.mas but any one will do. Select the Bulleted List or one of the other layouts. Type a title and bullet points in the appropriate boxes by first clicking on them to make them active; you can use the text in Figure 5-10 or make up your own.

Now that you have your basic page, you are going to add a short musical introduction. In order to change the "sound bite" for this and other presentations that use the same file in one operation, you will *link*, rather than *embed*, a sound file. Click on the Media Manager Smart Icon on the toolbar, indicated in Figure 5-10.

Step 2: Select, preview, and import a sound file

The Media Manager dialog box appears and displays your file directory. Search for and locate the Wave file you want. The multimedia objects available to the user via this dialog box are Wave (sound) files, MIDI files, and Movies (animations). You will find corporat.wav in the lotusapp/multimed directory.

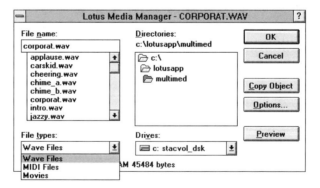

Select it and then click on Preview to hear what it sounds like (you will need a sound board and speakers).

Click on the Options button and specify in the Media Manager Options dialog box how the file is to be played.

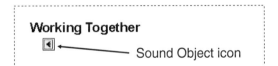

Select Play once in the Play Options box and select Refer to file in the File Options box, then click OK.

Close the Media Manager dialog box, and the screen returns you to your bulleted layout. Notice that the Sound Object icon appears on the page indicating that the Wave file has been linked.

Working Together

Sound Object icon

Step 3: Resize and move the sound file icon

Single-click on the Sound Object icon to select it. Handles appear, allowing you to resize and move it. Grab one corner and enlarge the icon. Then pull it to the lower-right side of your page:

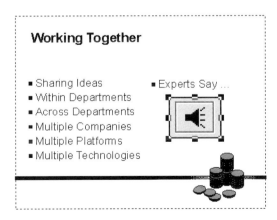

Step 4: Assign a behavior to the sound object

Now you want to determine how the sound object performs within the presentation. Use the right mouse button to access a Quick menu for the sound object (or pull down the Edit menu and select Lotus Media Object).

When the dialog box appears, select Play Options.

The Play Options dialog box allows you to choose when the object will play and how it will be triggered, as well as if the icon will display during the presentation.

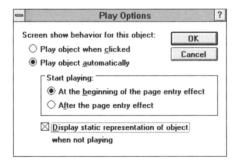

Select the options to play automatically and at the beginning of the page entry effect. Also, click on the checkbox to display a representation of the object. Close the dialog box by selecting OK.

Step 5: Play your presentation

You are now ready to run your screenshow and hear the sound play at the beginning. Click on the Screenshow Smart Icon. Other programs may refer to this command as Run or Start, but they all mean the same.

Media Integration: Macromedia Action!

Macromedia Action! is one of a new breed of multimedia presentation programs that focuses almost exclusively on creating presentations for electronic delivery. Its operating metaphor and user interface approach each individual segment of the presentation as scenes. The appearance, transition, and animation timing within the scenes are controlled via a timeline window.

The orchestration of multimedia events within the scene is far more precise and flexible than with more traditional presentation slideshow packages. Audio clips can be triggered in relation to visual images; video clips can be timed to coincide with text and other elements; animated objects can be coordinated with other actions as well.

Production with Action! begins with a template, a blank screen, or you can import a slideshow presentation from Freelance, PowerPoint, or Harvard Graphics. Scenes can be set to use a common backdrop, or they can have their own look. The final presentation consists of a series of scenes linked together.

In this tutorial, imagine you are the human resources director for the Farkles Corporation manufacturing company. You are scheduled to make a presentation to the management team showing them the positive effects of your employee morale program. You have used simple bulleted-text presentations in the past, but this time you want management to get a feel for your results, not just see facts and figures.

Step 1: Launch Action! and create a new file

Choose to Create a New Presentation in Action! from the startup screen. When the Template selection box appears, select Close; do not click OK. You will see a work area that looks more or less like other presentation software work areas, except for a timeline on the screen. If the timeline does not show, pull down the Windows menu and select the Timeline option. If you do not see a Tool palette, select that option from the Windows menu as well.

To select your background, pull down the Scene menu and click on Background. Select the Scene Options command, and a dialog box asks you what kind of background you want. For this demonstration, select Picture, then browse the clip media on your CD-ROM to find a suitable background. In the example (see Figure 5-11), a file that depicts an office building has been imported and fills the screen.

Step 2: Add music

To open with some music, click on the Sound Icon (a small speaker) at the lower left of the Tool palette. You are given the option of a Sound File or CD Audio. Select Sound File. The Sound File dialog box appears. Again, browse the CD-ROM and pick a track of music by highlighting it and clicking on OK. In this example, a recording called JAZZ has been selected. In the Edit Sound dialog box that is now on the screen, enable To End of Scene by clicking on the box, and Preview if desired. Click OK

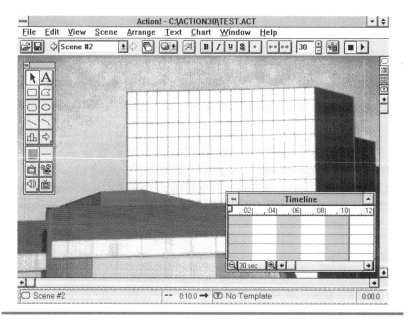

Figure 5-11   Import a background for the scene

to close the dialog box. Notice that the music track has been inserted at the top of the timeline and runs the length of the scene:

Step 3: Add three sets of text

To insert your headline and text, click on the Text tool ("A") in the Tool palette. The text entry box appears. Pick a color that will contrast with the background you have chosen by clicking on the Color Selection tool (colored square) in the Tool palette. Before or after you begin typing, you can pull down the Text menu and set the font, size, and style of the text. You should use at least 36 or 48 point for the size, and a sans serif font is best; bold is a good idea for headlines. Now type **The Farkles Corporation**. Click off the text, and the text appears superimposed with moving handles. Click and drag the text where you want it to appear. Use the handles to change the area and line breaks of the text, until it looks something like this:

Notice that the text has been added to the timeline.

Now so that you have elements to sequence, repeat the text creation steps two more times. Type **presents...** and position it on top of the first text. Type **Our Happy Family!** and position it on top of the other two. It will look like a jumble of letters at this point, but leave it as is.

Step 4: Add photos of employees

To add photos of employees, or any photos you choose from the CD-ROM clip art or another source, you first click on the Picture tool (the hanging picture frame) in the Tool palette. The Import Picture dialog box appears and lets you browse your directory for a file:

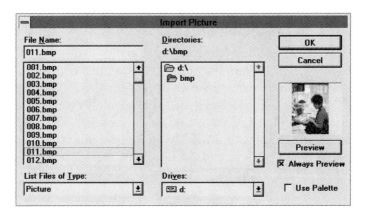

Find the picture that you want and highlight it. If you want to see the picture before importing, click Preview or select the Always Preview checkbox. Click OK when you have decided on an image. The photo is imported and appears full size on the screen. Use the handles to size it down and move it where you would like it on the screen. It does not matter if it covers the text.

Repeat the procedure for adding a picture, if you like. The more you add, the more practice you will have when setting the timing and sequence. Three pictures have been selected in this example. They have been sized and placed in roughly the position they should appear in the presentation.

Step 5: Set the timing and sequence of media elements

You should now have several text elements and several graphic elements on your screen together (some may be hidden by other elements, which is fine for now). Drag and adjust the size of your timeline so you can see all of the bars representing your media elements. Each bar extends the entire length of the scene, that is, from left to right. If you were to play the presentation now, everything would appear where it is, the music would play, and nothing would happen. Play it if you like now by clicking on the Play button (right arrow) at the top right of the menu bar or by pulling down the Window menu and enabling the Control Panel, then clicking on the Play button (right arrow).

To adjust the time when a media element appears, click on its bar in the timeline. You will know you have the right bar by the handles that appear on the media element in your scene. With the bar outlined or highlighted, move to the left edge of the bar and carefully grab the end by holding down the mouse button. Drag to the right, and the bar should get shorter. You may need a few tries to get the feel for it.

Leave the music track alone. Start with the headline text "The Farkles Corporation." Shorten it so it appears at about one second into the presentation, according to the time scale at the top of the timeline window. Now shorten the right end until the total length of the bar is only about three seconds.

Repeat this procedure with the "presents..." text, then the "Our Happy Family!" text. The exact length of each is not important, but you should set one to start as the previous one ends. Do this now for all the pictures you have selected. You might want these to overlap a bit for effect. Space the staggered picture tracks so they take up the entire length of the scene. You can, if you wish, change the length of the scene by dragging the vertical bar (at the far right) in the timeline window to the right (shorter) or left. When you have finished, the timeline should look roughly like this:

Step 6: Fine-tuning and adding transitions

Try the timing now if you like, by starting the presentation from the Control Panel. The music should play, followed every few seconds by a

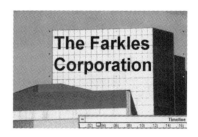

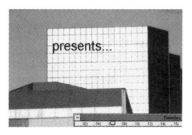

Figure 5-12 The final sequence, as created in the Macromedia Action! tutorial

new media element. At this point, you can begin to fine-tune your scene. Use the "scrubbing" handle that moves across the time scale (button and vertical line) to move your presentation forward and back.

You can also add transitions using the Motion Palette from the Window menu: Highlight the bar or track that represents the picture, and select whether you want an Entry or Exit transition. When the transition palette appears, click on the desired effect. After selecting your transitions, you may have to adjust your timing again. (Broken vertical lines in the timing bars represent the locations of your transitions.)

After you have your timing and positioning of all elements the way you want them, play the presentation and admire your handiwork. Figure 5-12 displays the sequence of images for the demonstration. To finish off the presentation, the "Our Happy Family!" text has been copied and pasted to the end of the sequence.

CHAPTER 6

Conversation is not a search for knowledge, but an endeavor at effect.
—John Keats

THE ELECTRONIC CONVERSATION: UNDERSTANDING INTERACTIVITY

nteractivity is multimedia's secret weapon and its ultimate promise. It works by linking both the goals of the presenter and the desires of the audience, letting each get exactly what they want from the experience. With the help of well-designed interactivity, you go beyond talking *at* your audience and begin talking *with* them.

The Power of Interaction

Human beings are the most interactive creatures on the planet. Our daily lives are a series of interactions with other people and the world around us. In the presence of another person, we listen, see, smell, and touch; gathering in information and considering it, we respond. Meanwhile, the other person is doing exactly the same thing. Each person processes the other's feedback, makes judgments and adjustments, and reacts accordingly.

You know what being trapped in a one-way conversation is like. The other person talks and talks but doesn't listen. Unable to make your point or get your questions answered, you begin to feel resentment, perhaps even anger. You begin to think about unrelated matters. Your mind wanders. If the talker is offering nothing of interest to you, or you question the validity of the information, you form negative opinions about the speaker.

Good conversation, on the other hand, is one of the most positive, useful, and pleasurable experiences in life. The hallmark of a "good" conversation is one in which both parties listen carefully to each other and respond with relevant information.

The same is true for a "good" presentation.

The power of interactivity resides in the ability of all parties to express their interests and communicate their concerns. For the presenter, interactivity presents an opportunity to learn about the audience, focus the message, and target specific needs on both an intellectual and emotional level. For the audience, interactivity presents an opportunity to get to the heart of what they came for. An opportunity to answer the question, "What's in it for me?"

Each person brings *individual priorities* to the presentation. A given audience for a presentation may include people with many different key questions they want answered. For example, in a business presentation for a product or service, some audience members may be concerned mostly about price. Others want to learn about comparative features and benefits. Still others need to know about availability, customer support, or quality standards. And, there are those who have come to check you out, to see what you are all about.

Multimedia technology offers you the means to address individual priorities much as you would in a conversation. When used to support a live presentation, multimedia allows fluid interaction with your audience and fast response to audience interests and questions. When used at a stand-alone terminal or in a kiosk, multimedia allows individual users to leap from topic to topic as new ideas or interests emerge, literally "shopping" for information as they go. Interactivity lets the presenter *operate*, not merely play back, the presentation in response to the concerns of a real (not abstract) audience in real time. It provides you and your audience access to multiple levels of information as needed and when needed—no waiting for "follow-up questions."

In effect, interactivity is multimedia's reason for being. It allows the user to create links between different media types and integrate them into a uniform multimedia document, program, or presentation. It is critical to understand that the links to the media and the information carried by the media are not fused and permanent. They are dynamic and active.

Interactivity, as it is employed in multimedia presentations, can be broken down into its four primary advantages:

1. *Addressing audience priorities:* Interactive multimedia presentations can be used to get to the information that the audience is interested in, or to help them find out what their interests are. Interactive presentations can also be used to direct your audience to interests you would like them to have. Most audiences are highly resistant to accepting a packaged spiel. Interactivity can give them a feeling of ownership and build credibility for the presenter.

2. *Allowing user-defined pacing:* When the audience is working hands-on with an interactive presentation, they can spend as much or as little time on a topic as they see fit. This allows them to fully absorb the material presented. The classic illustration of interactive pacing is in computer-based training (CBT), in which individuals move through the environment at their own pace, reading and answering questions at their learning level.

3. *Building audience associations:* People think associatively. When presented with a topic people tend to make connections and comparisons with related ideas or events (see the "Associative Thinking" section of this chapter). Interactive multimedia presentations help people build bridges between the topic at hand and their own experience. In this way, interactivity allows you to suggest to

your audience ways they might use a product, understand a concept, or grasp the significance of information.

4. *Integrating information:* Given the mass storage media now available, you can transform presentations from a simple set of bulleted slides into complete information environments. You can link text, graphics, sound, and video media from other presentations or databases to your presentation. This integrated information approach moves you toward a truly conversational presentation environment where the audience can interact dynamically and experience the subject on a deep and meaningful level.

Interactivity and the Personal Computer

Interaction is not unique to multimedia presentations. A good presenter knows how to "converse" with an audience—how to listen and to organize and access information the audience wants. Yet, the ability of a personal computer to store vast amounts of information and, more importantly, to access that information quickly and precisely in response to electronic queries, can endow a presenter with extraordinary powers of communication. Can it make everyone Super Presenter? No. Can it harness the power of the conversation and turn it to the presenter's advantage? Without a doubt.

The personal computer has been an interactive tool since its inception. CBT experts long ago proved that personal computers could make the learning process faster and more effective. In fact, many CBT programs were among the first to use digitized sound and graphics to enhance course materials and promote better communication. But until relatively recently, it was not practical to adapt these complex tools and techniques to presentations. That has all changed.

As the tools for digitizing text, graphics, video, animation, and audio have come down in price and gone up in ease of use, multimedia has arisen as the presentation superstar. The presentation and authoring software tools discussed in Chapter 5 have made interactivity accessible to every level of presenter, though not every level of presenter can create every level of interactivity.

Interactivity as it relates to personal computers refers to two aspects of the process. First, from the technology standpoint, interactivity describes the software links and "switches" between media files and visible objects on the screen—almost any visual element can be programmed to be interactive. The switches can be operated by the presenter, by the audience, or by the program itself. Interactivity also refers to the software programming or operations that make it happen. Second, from the communications standpoint, interactivity describes the design of the information or the structure of the content and how the interactive process will operate throughout.

Programming and operating the software can range from the simple act of creating a button in a program such as Compel and assigning it a jump action, to the complex use of scripting or a programming language inside a professional authoring environment such as Director. The challenge of designing the interactive structure, interface, and content covers a similar range, from basic links and jumps between a few main topics to intricate "information systems," which contain large media and text databases that can be accessed for presentation in real time. Creating an effective interactive presentation involves a solid vision of what is to be communicated to the audience, the technical tools to realize the vision, and a significant investment in time and effort.

The Cost of Digital Conversation

The emergence of interactive presentation technology gives you many more options, nuances, and techniques for communicating with your audience. But keep in mind, the added complexity of technology often comes with a correspondingly high price. You will encounter more complex, time-consuming, and expensive development and planning issues, as well as the added expense of the multimedia components and display media. As with most things in business, interactivity must be evaluated in terms of its return on investment. It must add significant business value to your presentation to offset the increase in costs and development time.

tip

Because the most basic levels of interactivity are the easiest to implement, involve the shortest development time, and use inexpensive tools, they offer the quickest return on investment.

The difficulty involved in determining the overall value of interactivity is due to its inherently subjective nature; knowing exactly what effect a conversation has had on the other person can be extremely difficult, if not impossible. A variety of evaluation tools are available, including the (dreaded) speaker evaluation form frequently handed out to the audience after a presentation, but no measure will tell you exactly how much more impact you are getting for your money with interactivity.

Nevertheless, the well-documented value of CBT, combined with the proven impact of multiple media on skills such as retention and learning speed (discussed in Chapter 2), make a powerful case for interactive multimedia presentation. Forgetting the empirical studies for a moment, simple intuition should also make clear that designing a presentation to be responsive to the needs, demographics, and interests of an audience increases the level of persuasion and, thus, the net return.

How We Think

Technology may change the format and tools used to gather and present information, but as already stated, it does not change the fundamental process of human communication. To make the best use of the new array of technical tools, you must begin by considering the way people process information and come to decisions. The extent to which you are successful in developing effective interactive presentations depends on your ability to use the techniques to address meaningful human interests and priorities.

Associative Thinking

Although you may have met one or two people who seem to speak and think in rigid outline format, most human beings do not think linearly. In fact, the ability to concentrate on a sequence of ideas one at a time, in order, with no digressions or jumps requires discipline and training.

By nature, the process of thinking is *associative*, a decidedly nonlinear process in which the mind makes an astonishing number of connections—some random, some controlled—to other ideas, experiences, or facts. In conversation, too, people are not driven by structured linear concerns but by feelings, instincts, and experience. People play off each other conversationally the way tennis players hit volleys—imagine how dull tennis would be if the players always knew exactly where the ball was going. As one person talks, the other makes creative leaps, inferences, and associations based on their own knowledge and experience. They then put their own spin on the subject in the form of a response.

As an extreme example, track the train of thought through the following conversation:

Bob: We saw a demonstration of a new presentation software program yesterday.

Jan: Darn, I missed the meeting. That new company from Seattle, right?

Bob: No, that funky little group from Carmel.

Jan: Oh, yeah. You know I used to live in Carmel.

Bob: Really? Is Clint Eastwood still mayor?

Jan: That was years ago.

Bob: I liked "In the Line of Fire." But I was surprised how old he looked.

Jan: There's a guy in the L.A. office who looks so much like him.

Bob: They all think they're movie stars down there.

Jan: They've got the best sales territory, that's all. It has nothing to do with talent.

Bob: Did you see the schlock presentation their regional manager gave at the dealer meeting?

Jan: I couldn't believe it. It was embarrassing!

Bob: Somebody should buy him a copy of that presentation software we saw at the demo. It lets you create interactive presentations that..

Interactive presentations give you the technical tools to communicate associatively. By opening up the presentation structure and giving people choices of how to experience it, you create the means for people to directly converse with you and your presentation. Choices and options let the audience travel in the direction of greatest interest rather than be herded in a single direction. As the user of the presentation reacts to priorities, interests, and questions, each layer of information builds a foundation for the next.

If you have been in business for any length of time, you have probably seen dozens or hundreds of noninteractive presentations in which the information is fed to you point by point, from beginning to end. Unless the presenter is taking questions along the way (or you happen to be sitting near the electric plug for the projector), you have absolutely no way to stop, redirect, or influence the course of the presentation.

By contrast, the book you are reading and the morning newspaper are, at least to some extent, interactive media. Why? Because you are the one that decides what you read and when you read it. You can jump ahead or go back. You can linger for a while on an idea. You can put down the book and pick up a dictionary or other reference material. In theory, you could even contact the author personally. The essential difference between noninteractive communication and interactive communication comes down to a question of control: who has control of the information and how it is delivered?

Traditional, linear business presentations are delivered beginning to end and designed with engaging "sell" points at the top, the "fine print" at the bottom. The objective is generally to convince people of the specific value(s) in a product, service, or concept. Almost all noninteractive communication conforms to a story model, that is, it flows from beginning to middle to end without being affected much by audience feedback.

With interactivity, by contrast, presentation is not locked into one track. The presenter can choose branches and jumps, or the audience can direct the connections. It is important to understand that, by giving your audience choices of the content they wish to see and the order in which they wish to see it, you are surrendering a certain degree of control.

For that reason, in interactive presentations, you must be extremely careful planning the structure or context of the information. You must be very clear on your objectives and see that the key ideas you want to advance are presented and repeated throughout the presentation. Keep in mind, interactivity means giving up some measure of control on how, when, and in what order the information is accessed and used by the audience, but it does not mean giving up control of the overall design. And paradoxically, good interactive design lets the presenter have greater control over the communication process by being able to draw on a potentially far larger content base.

Interactive Design

Designing an interactive presentation refers to more than just creating a look and feel for the production or what information will be delivered. It involves designing the experience that the audience will have as well. Interface design, a subset of the overall design and planning process, has become a recognized discipline in multimedia production. Interface designers are responsible for the clarity and appeal of the visual display of content. They ply their art to control the way interactivity is introduced and how interactivity is applied. The key to good interface design is making an interactive presentation easy to use and intuitively easy to understand, while furthering the communication strategy and objectives that underlie the project.

Storytelling

Most noninteractive presentations, regardless of the support technology, are structured as narrative. They follow an implicit storytelling model, in which a case is made by the presenter with the intention that the audience will adopt a point of view, assimilate the information, and/or take some specific action(s). As with any dramatic narrative, linear presentations have a defined beginning, middle, and end.

The goal of interactive design is to take people into the presentation, give them choices, and lead them to a result—just as in a traditional linear outline presentation. The difference is that you are giving yourself and your audience more choices and options to get to that endpoint and by a route designed specifically for them or by them. Think of yourself as a travel agent. It is their trip, but you are the one who will help them get there in the most meaningful way possible.

If the audience is not interested in one or more of the topics, the presenter risks losing their attention. If the audience feels that the structure of the presentation is designed to manipulate their actions, then they are constantly assessing the validity of what you are telling them. If each individual is not getting information that is responsive to his or her unique levels of interest, you lose both attention *and* credibility.

Any presentation must be written and designed. This requires time, effort, and a clear understanding of the specific objectives. Linear narrative presentations focus on building a case, beginning with audience interests and a base of facts, then presenting an argument, and finally ending with a solution. A simple sales presentation narrative might be:

▶ In your business you have a particular problem.

▶ You need a product that will solve your problem.

▶ We make that product and you should buy it.

In the design process, once the main and secondary points have been developed and put in order, the related graphics and other components that amplify and support the points are added. With multimedia, these can include almost any type of visual or audio. Once all the work is done, the presentation is complete, unless material is updated or deleted at some point.

Making Connections

Interactivity adds a new dimension to the design process. Not only will the presentation embody all the main points and secondary points, it will include connections and intersections to separate topics and ideas. Interactive design can offer a few or a few thousand choices and options.

With regard to content, remember, the more choices you offer, the more you risk losing your central points in the maze of the interactive web. If your goal is to focus an audience on a key result or action, such as buying a product, you will have to develop the content so that your themes and points of interest are threaded throughout the presentation.

Developing the media for an interactive presentation poses another tricky set of decisions. Moving in a nonlinear fashion from one connection to another will tend to juxtapose or group the various media types, disrupting the flow of the presentation. By chance a certain sequence of information, for example, might tend to be a long series of text slides with no video, animation, or audio support. Another series of connections might string together one video clip after another, with no text support. Design must take into account all the possible connections. For a very large presentation, the process can be daunting and can require sophisticated authoring and programming tools (see Chapter 5).

Updating an interactive presentation poses still further cost and time challenges. Changing the order of a few screens or media clips in a noninteractive presentation might be accomplished in minutes, whereas editing the interactive structure of a presentation can mean tearing apart the design and starting from scratch.

Gimmicks and Grandstanding

Before moving on, just a note about the use of multimedia as a gimmick. The relative ease of adding at least some level of interactivity to a presentation has tempted some companies to use the technique for the same reason mountain climbers climb mountains: because it's there. The ability to link and execute conditional operations is used more as a display of technological virtuosity than as a legitimate communication device.

This is fine if one of the goals of the presentation is to show off and dazzle the audience with the high-tech skills of the presenter; if not, it is a colossal waste of the audience's time and a violation of sound presentation practice. In any case, good interactive design should be transparent to the audience. With well-implemented interactivity, the audience is unaware of the depth or complexity of the operations that are occurring on the screen.

Your goal, as you become more proficient with using or evaluating interactivity as a communications tool, should be to understand when interactivity is working toward the purpose of the presentation and when it is being added like so much glitter onto an inherently deficient design. Offering the audience bells and whistles as a substitute for content is never advisable. A gimmick is a gimmick. A clearly structured content based on a clearly defined set of performance objectives and desired results has no substitute.

Interactive Structure and Strategies

The ways interactivity can be implemented are nearly limitless. Many high-end applications of interactive technology have been in the market for years, including sophisticated CBT programs, interactive video games, and touchscreen kiosks. The complexity and attendant economics of

Case Study:

Company: USX Corp., Pittsburgh, PA

Business: Steel manufacturing

Objectives: Develop an interactive multimedia system to teach employees about their retirement benefits using personalized, up-to-the-minute information.

USX Corp., one of the nation's largest steel companies, prides itself on the exceptional retirement benefits it provides its employees. But several years ago it realized that few employees understood the scope of the benefits and how to take full advantage of them. In response, USX developed inter-active multimedia kiosks for use as employee information centers.

Deployed at 13 sites around the world, the EASY Plan (Employee Access System for You) IBM PC-based kiosks connect directly to USX's mainframe computers at each site. Using graphics, animation, and sound to make the process under-standable and pleasant, the presen-tation guides users through the process of retirement planning via touchscreen. A graphical interface and audio feedback help them navi-gate through the system.

Because the kiosks are used by employees with varying levels of computer literacy, USX carefully de-signed the presentation to be simple and nonthreatening, says Bob McMaster, manager of benefits in-formation systems for the company. At the beginning, employees can choose to see a demonstration of how to use the touchscreen system. This presentation even illustrates the dif-ference between a "good" touch and a "bad" touch. To further facilitate user interaction, McMaster says they developed standards to ensure a con-sistent look and location for all of the navigational buttons. Each button also includes text to indicate exactly what it does.

"All of the buttons are very ob-vious, so people know exactly what to do," says McMaster. "We tried to keep it so simple that they wouldn't need instructions to learn how to use the system. The system is called EASY, and that's what we tried to make it."

As employees enter specific data, the program instantly com-putes their retirement savings, in-cluding benefits from their personal savings accounts, the company 401(k) plan, social security, and the company pension plan. Through the use of animated graphs, they see growth projections based on prevail-ing interest rates. They can also run projections showing how their retire-ment income will match different lifestyle scenarios.

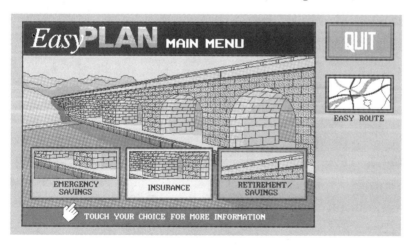

USX Corp.

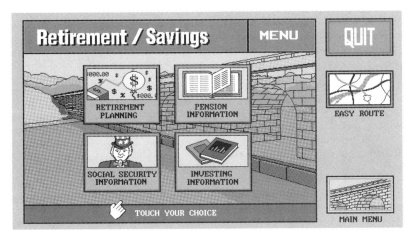

installed in 1989 and updated in 1992 with snappier sounds and visuals. According to McMaster, 85 percent of the company's salaried employees have used them, and most come back again for additional planning—some as often as once a month.

"Nobody wants to do these horrendous retirement calculations—and it can be embarrassing to discuss your personal finances with a personnel office," says McMaster. "These kiosks put our expertise in a box, while freeing the employees to model an ideal retirement plan." The kiosks have been so successful at USX that the company has decided to market the system to other corporations.

If they find their savings rate falls short of their desired goals, the program shows them how to take advantage of higher-yield savings mechanisms like stock funds or how to adjust their 401(k) contributions. As users make changes, they can immediately see the results. At the end of the session, a hard copy of their projections is printed and their work is saved on the USX main frames.

Created using IBM's IWPS authoring language and Storyboard software, the kiosks were originally

these high-end applications generally put them outside the realm of reality for traditional business presentations. This section focuses on three distinct design strategies you can use economically and effectively to create focused and successful interactive presentations with the available technology.

Level I: The Interactive Outline

Most basic linear presentations are broken into topic sections and subsections in an outline format and move along a linear path from the first heading to the last subsection. In a 35mm slideshow, the order of the slides in the carousel represents the order of the presenter's outline.

Unlike a slideshow, however, electronic presentation opens the possibility of random access to the slides, scenes, or images. Most presentation software packages offer at least some ability to jump from image to image. A few let you automatically jump to an entirely different application—a spreadsheet, a word processor, even another presentation program—during the presentation and back again when you have finished.

Using this basic level of link and jump interactivity, a presenter can move to relevant information, pass over unneeded sections, and import in real time most any media type. For example, suppose after running through an introduction and asking a few questions, you determine the next two sections of the presentation to be too basic for the experience level of the audience. Anticipating that situation, you have designed buttons or other controls that let you skip to the appropriate section. Within each section, the content is still organized in an effectively linear fashion, but now you are able to move from point to point, wherever links have been established. Some programs, such as Compel, automatically create "hyperlinks" to other slides or screens representing points in the outline (see Figure 6-1).

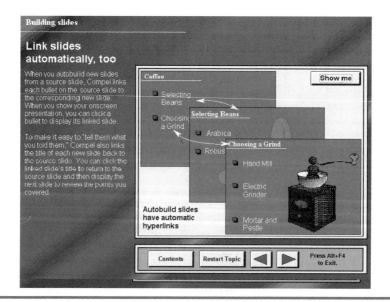

Figure 6-1   In some multimedia presentation software such as Compel, an interactive outline is generated as the slides are created

In effect, this relatively simple navigational ability—marking sections and creating nonlinear links—changes the presentation analogy from the slideshow model to the overhead transparencies model. Using overheads, the presenter has random access to transparency-based images. The images can be presented in any sequence as needed. Of course, working with more than a dozen or so transparencies can be extremely awkward. In addition, the images on a sheet of transparency film cannot contain hyperlinks; the presenter, not the program, must perform the search and find the next image.

Another way to look at this basic level of interactivity is as an electronic table of contents. As in a book or magazine, you can jump from the table of contents directly to the page where the information you are after is located. Adding action buttons to the table of contents in an onscreen environment adds interactivity to the presentation. The topic or heading serves as the unit of organization in the information structure and the interactive design. In Figure 6-2, the onscreen buttons offer a choice of three subsections, and an additional button exits from the presentation program.

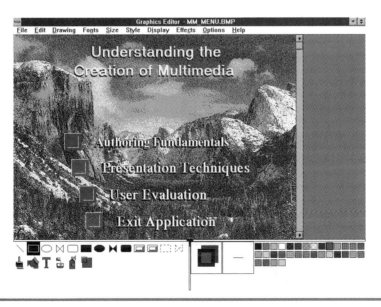

Figure 6-2 An interactive outline structure begins with buttons for each of the main subsections

Level 2: Hypermedia

Hypermedia is a term that was created to describe an interactive system or document. It gets its name from *hypertext,* an interactive system that was developed to allow links to be created in textual material. Hypermedia takes it a step further, to include not just text links, but links between screens and objects as well. A hypermedia presentation allows you to begin anywhere and move around within the application using *keywords* and common phrases, as well as objects and media elements that appear throughout.

In a presentation, this level of interactivity lets the presenter or audience jump between topics, subsections, and keywords from within the screen being displayed. It gives them the ability to follow key concepts, names, and phrases. It allows you and/or your audience to associate and make *cross-references.* Useful hypertext keywords or key phrases might include trade or proper names, places, dates, times, prices, and more. Figure 6-3 shows a hypothetical hypertext screen within a presentation. Clicking on any of the highlighted words or phrases would take the presentation to all other references to the same words or phrases.

Hypermedia allows a presentation to deliver information with a thematic thread by tying together every reference of the themes that the presenter wishes to convey. A particular product, technology, or feature, for example, could be established as the central theme.

Another way to look at hypermedia is as an electronic index. Where the electronic outline discussed in the last section uses the headings as the unit of interactivity, the electronic index uses the keyword or key concept.

Hypermedia also allows you to combine keywords to find sections of a document that contain information in which two or more keywords are active. In Figure 6-3, you could, for example, select "advertising schedule" and "Boston" to jump to the specific plans for that city.

Hypermedia presentations are relatively easy to assemble, using the current generation of software. You could, for example, assemble a robust hypermedia presentation by selecting terms, trade names, and proper names to act as keywords. Figure 6-4 shows the dialog box in Astound that lets you assign a link to any object onscreen, including text.

A distinction must be made between linking words each time they appear and programs that will actively search an entire presentation or database for each occurrence. The former allows you to jump to a screen

> ## PRODUCT STRATEGY
>
> In this presentation you will learn about Product1 and Product2. They are intended for use in the medical and cosmetics industries, as well as by consumers interested in personal hygiene. Key accounts include LookWell Corp. and Health Habits, Inc. The advertising schedule will begin in May 1995 and the products will hit the market in June 1995. The initial roll out will be in Los Angeles, Boston, Miami and Chicago.

Figure 6-3   An example of hypertext as it might be used in a presentation. The highlighted words contain hypertext links to related topics

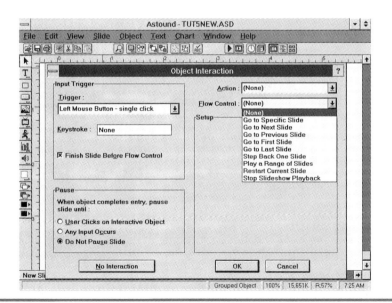

Figure 6-4   In Astound, any object on the screen, such as a button or text, can be linked to any other screen

containing the keyword. The latter initiates a programmed search for occurrences of the word or phrase and requires scripting or programming.

Level 3: Conditional Branching

So far, interactivity has been limited to single words, topics, or phrases. In the first two levels, interactivity can be achieved without the need for a high-level authoring system, scripting language, or programming. As interactivity becomes more complex, however, the simple tools of jumps, links, and hypermedia meet with limitations. With jumps and links, for example, branching to topics, subtopics, and sub-subtopics is possible, but at some point the screen is cluttered with option buttons, and the links to other information become inaccessible.

The third level, here called conditional branching, expands on links, jumps, and simple branching. At this level, the information is organized and the system is programmed to perform operations that help the user (presenter or audience) find deeper and deeper levels of information. It also lets them find more specific information in a much broader database than a presentation with simple interactivity. In effect, the user is operating the information system as the computer program takes and fills orders.

The idea is to send the user on a journey that has a large number of options for learning about any topic that is available and to let them quickly reach the level of detail they desire. Branches can lead to other databases, other presentations, even onto interactive networks. The layers of information can be as dense and detailed as the designer wishes to attempt (see Figure 6-5).

To get an idea of how conditional branching works, take a look at the Help system for most software packages on Windows and Macintosh machines. As you select a feature, the feature either opens immediately or displays a list of available options in a dialog box. The dialog box choices serve as prompts, to suggest and remind you of all the ways you have to use the feature. The program asks you to input your requests and responses, then goes to the correct information, performing whatever preprogrammed operations are necessary to get you there. The same concept is at work in a multimedia kiosk or a CBT system. The key difference is that the user is operating the presentation system, not merely playing a presentation. Every person who goes through the conditional branching presentation will create a unique information event, a custom conversation.

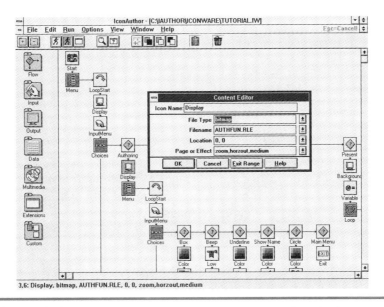

Figure 6-5 A conditional branching structure in IconAuthor includes programmed or scripted operations that are run when an event is triggered

Structure Maps

Building even the simplest jumping, linking, or branching interactive structure, as opposed to a linear one, means designing a framework for cross-references and links among different topics and subtopics. You will discover that developing and planning interactivity involves a broader three-dimensional planning and thinking approach than you may be accustomed to with simple outlines. The more you experiment with this type of organization, the more effectively you can use the available multimedia presentation and authoring tools.

When you sit down to write a letter, report, or other business document, do you first take a moment to outline your material? The outline serves as a map of your content. It provides the boundaries and subdivisions, the routes and the avenues to and from information. Unlike the basic outline structure you were taught in grammar school, which steps methodically from one point to the next, the outline you design for interactive presentations must consider three levels—simultaneously:

▶ The content you want to present

▶ The links and cross-references you want to make

▶ The action or result you wish to encourage

An *interactive structure map* (a graphical representation of the presentation's framework) is an effective way to identify and incorporate these three interactive structural elements. Mapping gives your information and its interrelationships a visual structure that is far easier to work with than a simple text outline. It lets you see where interactive features are needed and where they are unwelcome. It also helps you visualize how you would like to break down your topics and the types of links between content points you would like to provide to your audience.

One useful technique for understanding the concepts of mapping is to think of your presentation in terms of one of those airline route maps you find in the back pages of in-flight magazines. Generally, an airline will have one or a number of hub cities that control the air traffic in their area. So a hub will serve a number of its own destination cities, while also having service to other hub cities. You can think of these hub cities as the places where your key topics come together and the routes as the links connecting them. In turn, each topic may branch into several of its own subtopics, just as an airline hub may serve several local flights.

Mapping Tools

Once you begin to craft a visual structure, you will begin to see creative ways to make your information come alive and help your audience achieve a richer sense of your content through interactive connections. There is a variety of ways to put your structure on paper. You can, of course, do your mapping within a high-end authoring tool as depicted in Figure 6-5. Some designers use hand-drawn storyboards or flow charts (see Chapter 4). Still others use project management software, database software, or org. charts. The mapping system you use will be largely a matter of personal choice.

What follows are seven mapping elements you need to lay out and structure your content. Note that the list begins with the starting point and concludes with an endpoint.

1. *The starting point* is where the audience enters the presentation. Depending on your content, you may want everyone to start at a single point of entry so you can make an introduction, establish terminology, and make central value statements.

   If you are dealing with a product that appeals to a variety of individuals in different markets, you might want multiple points of entry. Using multiple points of entry lets individuals in your key markets go directly to information of specific interest. Not incidentally, you will let them know that you take their marketplace seriously and are working to meet their needs. In addition, by letting the audience choose the point of entry, you are increasing the shelf life of the presentation; the same presentation works for multiple audiences.

2. *Hot buttons* are the controls the audience or the presenter uses to activate links to other screens, other topics, keywords, or media events. They are, in a sense, the tickets that get you on the plane at one place and off at another.

3. *Links* are the lines between two hot buttons. They establish a relationship between topics in the presentation and allow your audience to move through the presentation content based on their own interests and associations.

4. *Topic headings* are the uppermost level of individual information blocks. They can be used as starting points for the entire presentation or they can be selected as you go, depending on the level of interactivity you have built in.

5. *Keywords* are programmed words or other elements that immediately jump the presentation to other occurrences of the same keyword. Keywords let users cross-reference the content based on specific terms or features. They can be combined for more detailed searching.

6. *Story vectors* are the series of links in sequence that establish a pathway through the presentation to a result. As users go through

the presentation, hitting hot buttons, selecting keywords, exploring links, and actively choosing options, they create a path of cross-references between topics. Your design should encourage and help users to find the key story vectors you think are desirable. Different individuals with different interests may create dramatically different vectors.

7. *Endpoints* are analogous to hub cities in the airline route map concept. These represent the "sell points," the "take away" points, and the "key benefits" that your presentation is driving at. You should be sure to identify these places early on and make it easy throughout your presentation to get to these endpoints. As with the airline model, most if not all cities have a direct link to the hub city.

Value of Mapping

As you develop your structure map and select your starting points, hot buttons, links, topic headings, keywords, story vectors, and endpoints, the level of interactivity that you need will begin to take shape. The more complex the interactivity—the larger the map and the more links—the more design and development time will be required.

The interactive structure map is a means for you to clarify your own mission, to pick out what is important for the presentation to accomplish versus what is merely nice or a bit interesting. Designing from an interactive story map, you will be able to visualize the whole presentation taking shape. It will give you greater control over the planning and design stages. Perhaps the most important value in the mapping process will be the down-the-line cost savings. As with any project, and more so with multimedia production, thorough planning will result in more effective use of time and resources (see Chapter 4).

Interactive Environments

A final aspect of your interactive design—a consideration you should weigh when selecting your level of interactivity—is how the presentation will be used, in what surroundings, and by which individuals.

Electronic media makes it possible for you to use the same development effort to create the same presentation in several different formats:

▶ *Speaker support:* An interactive presentation operated from a portable or desktop computer provides the presenter with the means to respond more directly and intimately with the audience. As questions arise, the speaker can move between topics and engage the audience interest and response.

▶ *Guided group discussion:* For informational presentations, interactivity offers, among other benefits, the ability to keep the main topic in focus while exploring related information on demand. The discussion facilitator can run a core presentation from a computer to stimulate discussion and new ideas and, when needed, jump to supporting media elements.

▶ *Executive information:* Elaborate "executive information systems" can be developed that feature video, sound, text, graphics, and live links to business data. At the top management level, this degree of interactivity gives the user access to more information, faster and on a more timely basis. Its benefits include the ability of managers and executives to make informed decisions more quickly and the capability to work with information in multiple media types—an invaluable aid to comprehension.

▶ *Computer-based training:* In educational applications, the interactive presentation functions as a virtual teacher. The actual instructor can be present or absent. CBT can also track student responses and use the information to customize the presentation for the student, making corrections or providing reinforcement where necessary.

▶ *Kiosks:* Interactive, audience-driven kiosks can be designed to deliver and gather information in a variety of public venues and private environments. They can be self-contained or they can function as interactive terminals connected to a central computer. Kiosks in a retail environment can be used both as a customer information system and as a training tool for employees.

Interactive Information Systems

The economic logic behind interactive design ultimately leads toward large-scale and high-productivity "information systems," in which presentations become an integral component (see Chapter 1). If you imagine an interactive presentation as a computer program running on top of a multimedia database or network, it takes on much larger implications for the communication process.

An interactive information system links the presentation to all or some of the resources in the system. Possibilities include accessing real-time market data, product sales, shipping dates, stock levels, receiving dates, rate tables—in short, any of the specific data necessary to satisfy the needs of a customer/audience at the time of the presentation. It could eliminate the sale-busting phrase, "I'll have to check that and get back to you." This type of system can be delivered from the desktop or can even run on a high-powered multimedia laptop. A number of companies have systems of this type up and running. (See Case Study in Chapter 1.)

The two most obvious advantages of an integrated information system are higher productivity and better communication. Another advantage is the ability to eliminate or reduce the creation of multiple types of presentation or sales media—brochures, overheads, videotapes, and handouts can all be melded into the interactive information system, complete with an option allowing you to print anything that appears on screen. Using periodic data updates over a network connection, or via modem connections from the field, you can maintain the currency of data in an information system. The overriding benefit of this approach is an immediacy and flexibility in the way your company presents information.

Eventually, nearly every company of any size will have some form of an interactive multimedia information system. Presentations will cease to be entities separate from the data source. Their purpose will go far beyond packaging the information in an appealing and persuasive form. They will truly have the ability to conduct an electronic conversation.

CHAPTER 7

Pay attention to the form, emotion
will come spontaneously to inhabit it.
—André Gide

SENSUAL PERSUASION: UNDERSTANDING MEDIA AESTHETICS

Multimedia by its very definition is about engaging the senses. When artfully employed, it allows you to reach deep into the emotions, down into the gut where most individuals ultimately sign off on decisions. To use media intelligently it is necessary to go beyond reason—to understand how pictures, words, sounds and moving images touch your audience.

Media Mastery

Successful presentation involves more than organizing details and outlining points. To be effective, you must *move* an audience. You do this not with facts but with feelings—the feelings you convey with words, images, video, and sound. Many presenters fail to realize that they are at all times communicating on multiple levels of consciousness. Presentations contain meanings that are both denotative and connotative—concrete and interpretive. Understanding the subtle influences of aesthetics is critical to the success of any presentation, but more so for multimedia.

We are sensual beings. Our world view, our perception of self, and of existence, are literally built from sensory input. It stands to reason that by manipulating the senses of an audience we can subtly guide them in a desired direction.

When it comes to aesthetics there are no hard and fast rules. But there are guidelines of good taste. And there is well-documented research establishing the inherent power of text, sound, and visuals—working separately and together—to influence such subjective elements as mood, attitude, and perception.

As you read through the following list of multimedia's uses, begin to consider how each of these tasks can be accomplished on a subconscious as well as conscious level:

▶ Increase audience awareness and attention

▶ Heighten and focus excitement and enthusiasm

▶ Explain abstract and complex ideas

▶ Communicate with people of varied backgrounds and learning styles

▶ Create simulations and reenactments

▶ Motivate individuals to take a course of action

▶ Engender positive feelings about you and your message

Levels of Communication

Information of all types reaches an audience on two levels—*conscious* and *subconscious*. On a conscious level your audience will notice when a graphic design is well balanced and clear. Illustrations clarify and draw attention to points being made in the text or narration. Sound effects and music alert an audience to the point that is being made and can indicate when an action is to be taken. Video and animation describe events and content in a time frame your audience will recognize and relate to. On a conscious level of communication, the prevailing rule is to use a specific technique only if it helps you to make your point more clearly and more powerfully than without it.

On a subconscious level, the rules are slightly different. You may choose to use an element in your presentation even though it will have no overt effect on your audience—they may not even notice it. In this instance, your goal is to design elements that trigger specific emotions, not ideas. You want your audience to form opinions and associations with the presentation message, but they will not necessarily make a conscious connection between the aesthetic design elements and your information.

Talent Assessment

You do not have to be an artistic genius to create effectual presentations. Even a modest talent for design will serve you well when it comes time to adding color, illustration, or photographs. Likewise, though you may not have formal training, you can teach yourself to be a good judge of sound, music, and video. It comes down to knowing the basics and trusting your instincts. On the other hand, you may have true artistic talents in one area or another, in which case you will likely find multimedia production to be a challenging and fulfilling outlet as you use your talents to enhance your business communications.

Learn to play to your strengths. Experiment with the different media types during your free time and discover where your talents reside. Make an honest assessment. If you have the talent, fine. If you do not, then admit that you will need the help of an artist or multimedia specialist (see Chapter 12). A skilled graphic artist, for example, can make the difference between communication and confusion. A talented musician could be the key to setting just the mood you are after. And, of course, the arts of photography and video can be powerful beyond all expectation in the hands of a professional, but also have the capacity to tarnish an otherwise competent presentation when poorly executed. This chapter will help you understand some of the important aesthetic concepts, which will enable you to better judge the aesthetic appeal of others' work as well as your own.

Media Selection

There are different schools of thought for graphic design, music, animation, and video as they relate to multimedia presentation. Selecting the right media is both an objective and subjective process. Table 10-1 can be used as a general guideline for assessing the relative strengths and attributes of each type.

Graphic Design

Graphics are the design glue that hold any visual presentation together. They can range from the simplest use of line and color to elaborate masterpieces, but there is one cardinal rule: The essence of good presentation graphics—or good presentations for that matter—is knowing what *not* to include, or in the words of Robert Browning, "Less is more." Uncluttered screens make the point more clearly and power-fully, and a few apt, quality visuals will do more for your message than a legion of mediocre ones.

You can use graphics to lend a consistent theme to your presentation as well as to demarcate key topics and subsections. They can be used to make important ideas leap off the screen and relegate less significant details to the background. Your design selections establish the image and tone you are trying to project and, by virtue of how well you select and use them, graphics will suggest to your audience the degree of profession-alism and attention to detail you have invested in the presentation.

Layout

Onscreen and projected presentations are limited by the rectangle of the computer screen. In any design screen, real estate is at a premium and should never be used carelessly. Good layout means proper use of the screen space, and therefore good graphical communication. Multi-media complicates the task of effective layout design by introducing such unique elements as video windows and navigation controls—buttons and

MEDIA TYPE	ATTRIBUTES
Text and narration	Written words are descriptive, detailed, and direct. They can be literal or evocative or both. Careful use of words is critical—they are easily misinterpreted. Narration can be informative, as well as highly expressive. Use words to say what you mean, but use them with the voice of conviction.
Graphics and illustration	Designs, drawings, and paintings can be used thematically, literally, or symbolically. Graphics can be explanatory, conceptual, or suggestive. They can be customized to the information, and targeted to the audience. Use color, style, and design to create a mood or atmosphere. The graphic design connects the disparate elements of the presentation.
Still photographs	Visually rich, detailed, and attention-getting, stills can convey realistic images and information. They can be highly suggestive, even symbolic. Artistic photography will grab an audience. Dramatic subject matter makes both a literal and subtextual impact.
Charts and graphs	Ideal for data visualization and comparative studies. Numerous variations on types and forms. Can incorporate some thematic and conceptual elements, but tend to be literal in nature. 3-D and creative thematic elements can alert a complacent or jaded audience.
Video and animation	Highly realistic and descriptive, they can also be highly entertaining. Use video to communicate time-based information in a time-based manner. Motion grabs attention and has a wide artistic range. Animation can be literally descriptive or suggestive. Works well for explaining or clarifying complexities.
Sound effects	Add audible texture to visuals. Provide audible cues. Emphasize points. Add entertainment value. Reach audience on a mostly subconscious level.
Music	Sets mood and tone of the presentation. Evokes feelings in the audience. Highly expressive, engrossing, and entertaining. The most subtextual of all media. Use of music can be unpredictable. Use with caution, but do not resort to blandness.

Table 7-1   General Guidelines for Choosing the Right Media

icons. Animation, too, changes the dynamic relationship of elements on the screen as it progresses, further challenging the talents of the designer.

Good graphical layout often involves creating a *grid* that defines separate spaces or regions for text, graphics, windows, and program controls. A strong consistent layout guides the viewers' eyes and gives them visual cues as to how the presentation is organized. The audience learns to expect where information will appear and how it fits into the overall scheme. Figure 7-1 shows a sample grid that could be used to guide the layout of the entire presentation. Note that each element in the presentation is assigned its own, familiar location where it will reappear in subsequent screens. The buttons marked "global controls" would always be present to keep the user oriented and give him or her control. For example there might be one button to return to the main menu, and one to exit the presentation.

One of your main design goals is to achieve a *balance* of the various elements on the screen: the text, graphics, windows, gutters (space around edges), and negative space (space between objects). Be aware that balance does not mean that objects must be symmetrical on the screen. In fact, symmetrical designs are frowned upon in most situations

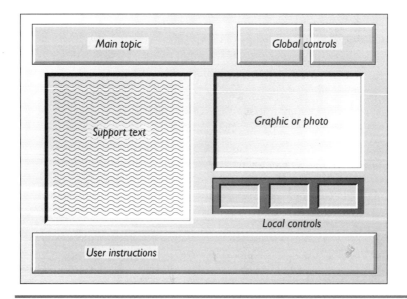

Figure 7-1   An "invisible" grid guides placement of design elements in the layout process

for their lack of visual interest. Balancing a layout means juxtaposing elements in such a way that the screen does not feel too heavy in any one area. Figure 7-2 demonstrates a well-balanced layout in which the four rectangular navigation buttons balance the rectangular shopping bag.

Designers frequently rely on what is referred to as the *rule of thirds* to keep layouts visually interesting. Rather than aligning objects on the center vertical or horizontal axis, the screen is divided into equal thirds in both directions. Images that are offset from center in either direction have more visual interest. The use of thirds creates better dynamics between the object and the borders of the frame. With more proportional relationships to consider, the eye naturally spends more time taking in the scene. Images set in thirds also seem to move or lean toward the central axis. Figure 7-3 depicts a building aligned first on the center horizontal axis, then on the lower third. Note how the first illustration looks awkward and out of balance. Moving it down so that it fills the center third creates a balance that the eye finds appealing.

Design theories are full of exceptions. One of these is referred to as the *power of centers*. This layout works like an archer's target, drawing the

Figure 7-2   A well-balanced layout (courtesy of Medior, Inc., San Mateo, CA)

Figure 7-3   Dividing the layouts into thirds adds rhythm to the design while maintaining the spatial balance

viewer's focus to the bull's eye. Although this may appear at first to be in violation of the rule of thirds, there are certain times when the best place to put the key element is dead center. Specifically, centered layouts are often used for large images that would throw the screen out of balance if put to one side or one corner (see Figure 7-4).

Figure 7-4   Sometimes centering adds power and focus to a layout (courtesy of Hypermedia Group, Emeryville, CA)

Within the context of your grid, you are not limited to north/south, east/west orientation for your graphical elements. *Diagonal lines* or compositions can be used to add interest and attract attention on the screen. Diagonals can also suggest a direction for the viewer to follow. Perceptual studies have shown that photos or illustrations set at a slight angle on the screen hold the gaze of the viewer for a longer period of time than the same elements placed on the horizontal. Note in Figure 7-2 how diagonal beams of light were used behind the shopping bag to add interest.

Evaluating Your Layout

Take a look at Figure 7-5. Notice the way your eye naturally follows the diagonal lines created by the guitar, the background text, and the background texture. Your eye is not allowed to roam off the page or away from the text. As you look down the left side of the image and reach the bottom, the large bold letters lift your gaze up and to the right. The guitar then guides you higher and to the left again. Notice also how the lines

Figure 7-5   Powerful composition in this screen guides the viewer's eyes to key visual reference points (© 1994 Ikonic Interactive, San Francisco, CA)

formed by the graphic elements come together to form a triangular shape that keeps you focused on the center of the screen.

Other elements you will find in this and any image with strong layout aesthetics include the repetition of graphic elements, parallel lines, converging lines, curves, coloring, texture, and shading. Practice evaluating layouts as you look at the many screen shots you will find in this book. After a while, recognizing the elements of good layout will become second nature.

Visual Clarity

The goal of good visual design is clarity. The moment your audience looks at the screen it should be obvious to them both consciously and subconsciously what is important and what is not. Too many bulleted items, too many icons, too many buttons, too many photographs—even too many colors can clutter and confuse the object of your message. As a rule, each separate image or sequence in a multimedia presentation should have only one main point and no more than three subpoints. This rule applies to text and to graphic elements. The image shown in Figure 7-6 demonstrates how a layout can include many elements and yet retain visual clarity.

It is common in poor designs for information to be clouded and distorted by competition between onscreen elements because the presenter tried to accomplish too much at one time. The unwanted interference results from the way such elements as shape, texture, line, line weight and color compete for the viewer's attention and the way they sometimes negate each others' purpose. When clashing elements obscure the content and control options they cease to be desirable, even if each element individually is a work of art. If your central element fails to stand out, the screen is probably too crowded or the design is in conflict.

Nonessential elements such as borders around text, charts, video windows, and buttons should not be more interesting than what they contain. Strong color and heavy borders can do battle on the screen and generate *spontaneous shapes* in the negative space between them. High

Figure 7-6   The many screen elements in this screen work together to communicate a great deal of information (courtesy Medior, Inc., San Mateo, CA)

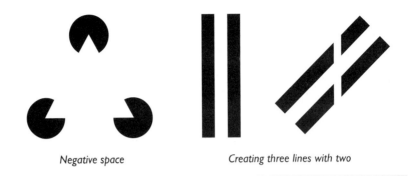

Negative space   Creating three lines with two

Figure 7-7   Unintended shapes can be formed by negative space

contrast regions, too, can create new shapes in the negative space between them. Figure 7-7 demonstrates how unintended shapes can appear. Figure 7-8 shows how the contrast between dark lines around buttons creates an unwanted grid of white lines between. Keep in mind that good information can be buried by bad design. On the flip side, it is also a fact that good design cannot do much to salvage bad information.

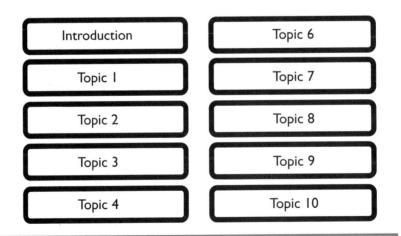

Figure 7-8   Heavy border lines, evenly spaced, create an
unwanted white grid

Backgrounds and Textures

Backgrounds serve as the canvas on which you will splash your text, illustrations, and other important presentation elements. From an aesthetics point of view, backgrounds can send an instant message about your presentation because they often appear briefly before any other elements and tend to be the first thing an audience notices, even though it may be subconscious awareness.

As with color, discussed later in this chapter, textures can have connotations that extend well beyond their literal association. The emotional message behind a texture can be generated by its tactile qualities or by social and cultural attitudes. Use textures in your backgrounds to subtly underscore your message. The following are a few examples of evocative textures and backgrounds:

- *Leather:* Suggests elegance or craftsmanship—a hallmark of luxury cars and fine furniture

- *Marble:* Embodies a cool and solid feel that translates as classical, reliable, and lasting—a symbol common to financial institutions

- *Handmade papers:* Suggestive of high refinement, artistry, and the days before automation—used to highlight the importance of documents

- *Gemstones:* Obvious connotations of wealth and elegance, sometimes hardness or permanence or special qualities—a common symbol of businesses catering to high-end consumers

- *Industrial metal plating:* A modern symbol of no-nonsense ruggedness and reliability—used as a background for presentations in the construction and manufacturing field. Also used by the computer industry as a counterpoise to the delicate and dust-free nature of microelectronics

- *Blue sky:* Suggests freedom, space, and potential—used often in advertisements to disconnect the product from the mundane everyday world and suggest wide-open possibilities (see Figure 7-9).

Figure 7-9 A background of blue sky and white clouds suggests a limitless potential for the vendors' high-tech products in this multimedia presentation (courtesy of HyperMedia Group, Emeryville, CA)

Thematic Design

The use of theme or metaphor to create a connection between elements in a presentation can be more than just a way to integrate buttons, backgrounds, borders, and the like. By selecting objects for repetition in the presentation you can use the associations of the object(s) to trigger a memory or engender a feeling, perception, or mood in the audience. Going one step further, graphics can be employed to create an atmosphere or milieu designed to engage the audience by virtue of the association that is common to the images.

A common example of thematic graphics is the use of Greek columns and designs to suggest a classical feeling. Presentations often pick up on some current cultural theme—Indiana Jones the archeologist, for example—to tell a story and create a mood. Refer to Figure 7-10 to see an example of how graphics elements can be used to draw the audience into the story and put them in the right frame of mind for the message.

Figure 7-10   Compaq Computer used a detective theme to tie
together its message and draw the audience into
the story. (©Compaq Computer Corp. 1992.
courtesy of the Hypermedia Group, Inc.,
Emeryville, CA)

Case Study:

Company: Alcon Laboratories, Inc., Fort Worth, Texas

Business: International ophthalmic equipment and solutions manufacturer

Objective: To create a series of interactive sales/informational titles for use in the company's trade show booth, traveling sales presentations, and sales training.

Alcon Laboratories, the largest international ophthalmic equipment and solutions manufacturer, has an eye for detail. So when it designed its series of six CD-I-based interactive presentations for its trade show booth it used high-resolution, 24-bit color photographs of Alcon's ophthalmic products embossed on rich, granite backgrounds. To enhance the effect, the granite background was also used in the booth design. In addition, the navigational buttons on the screen mimic the actual control buttons of the company's flagship product.

"The goal was to create touch-screen applications that looked like they were all part of the design of the booth and the product lines," says Wes Wright, a partner in Inmedia, Inc., the Dallas interactive training and presentation firm that created the series for Alcon.

Alcon supplied Inmedia with actual samples from the booth, which Inmedia matched with a stock texture photograph and scanned into Adobe Photoshop for Windows running on a Dell 486 DX2/66 computer. Alcon also supplied the production company with slides of the products they wanted to incorporate. The slides were digitized to Kodak Photo CD format and ported to Photoshop as well. To create the navigational buttons, Inmedia used photographs of the control buttons on the Alcon product as models when recreating them using Autodesk 3D Studio modeling software. The entire project was authored using OptImage's Media-Mogul software on a CD-I authoring system.

But the beauty of these programs is more than just skin deep. In order to allow users of the presentation to explore the features and benefits of Alcon's products in depth, including viewing actual surgical and clinical demonstrations, more than 1½ hours of professionally produced, full-motion video was incorporated into the six disks. The extensive video clips were shot using Sony Betacam SP. Inmedia sent the tape to Advanced Multimedia, Inc., to encode it to MPEG. The digitized video was then sent back to Inmedia to be incorporated into the CD-I program.

While the rates charged for the compression process have begun to go down from their early high levels, it is still expensive, costing anywhere from $80 to $250 per finished minute of video (depending on the source material, the quality needed and the volume). Even so, Alcon plans to add more video to its existing titles to take advantage of video's power to demonstrate, says Wright. Currently, the physicians' testimonials in the programs are only in text form. Wright says the long-range plan calls for shooting the testimonials on video.

Alcon Labs

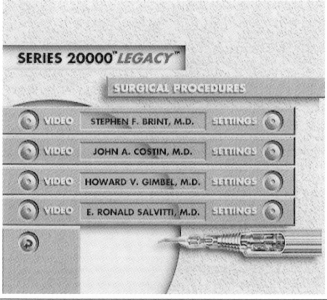

Each of the CD-I presentations also has a vibrant narration track recorded by a professional voice-over specialist as well as still photographs, graphics, and text. The overall effect is a rich, visually detailed environment.

An attractor screen shows a product revolving in a window. The user activates the program by touching anywhere on the screen. This brings up the main menu, which allows the user to choose from three to five topic areas, including features and benefits, demonstrations, and physician comments.

While watching video clips, viewers can use a touch-screen control bar placed on the bottom of the screen to pause the video, play in slow motion or frame by frame, continue normal play, stop, and return to menu. The frame-by-frame controls have proven particularly popular with physicians who can see how to perform a procedure step by step. Many actually sit at the kiosks taking notes, says Wright.

Color and Perception

Color is one of the most subjective elements in the presentation design process. Most of us understand the basic uses of color—for example, bright colors draw attention and muted colors blend in. Its effects can be extremely subtle or overt. The use of color has probably received more study and attention than any other element of design (see Table 7-2). Yet, the power of color and its ability to influence our emotions, perceptions, moods, and actions is little understood by most people.

A background in the fundamentals of color theory and how it is used in presentations is useful for creating effective designs. Color will be addressed here as an element of graphics design, but it should be noted that the principles of color usage apply to all aspects of visual media, including video and animation.

Color has three basic uses in presentations: *identifying*, *contrasting*, and *highlighting*. You might, for example, identify all fourth-quarter sales by making those figures blue; contrast profits and losses with black and red; or highlight a key point with a bright orange bullet. Color can be used in backgrounds, text, lines and shapes. It can be used in patterns, textures, and gradients. Of course, color also can be used in full-color, photorealistic images or video.

FOR ADULTS	FOR CHILDREN
Blue	Yellow
Red	White
Green	Pink
White	Red
Pink	Orange
Purple	Blue
Orange	Green
Yellow	Purple
1Source: Symbol Sourcebook, Henry Dreyfuss, McGraw-Hill, 1972, NY, NY	

Table 7-2   Most Popular Colors (In Order of Preferences)

Color, as most everyone instinctively understands, can both attract and repel. Bright, vivid colors and colors that contrast sharply with each other grab the eye. But they may produce unwanted negative effects in the process. Bright or busy colors on a panic button (or a fire hydrant) may serve the purpose of alerting the user, but the same colors in the background of a text slide will tend to pull the eye away from important content. Complementary colors such as green and red may serve to trigger images of the holiday season, but they can also fight with each other by creating an optical strobing effect when placed close together and stared at for too long.

As a rule, reds, yellows, and oranges will seem to project forward and dominate the screen while purples, blues, and greens will recede. Similarly, shapes painted in bright colors will tend to look larger to the eye than darker-colored shapes. Try using your computer to draw a simple box on the screen using a white or other neutral-colored background. Duplicate the box. Then fill one with a bright orange and the other with a dark blue. Notice that the orange box appears larger. If you were to design control buttons in contrasting bright and dark colors you would run the risk that your design would appear out of balance. The implications also are obvious for treating text or other screen elements. All elements on the screen that relate to the message should be given a color treatment appropriate to their level of importance. Conversely, you should avoid diminishing elements by improper use of color.

Color Gradients

Gradients, gradual color transitions that can be used to give the illusion of spatial depth, are one of the most popular color devices in computer design. Gradient techniques are often used to give 2-D objects a 3-D appearance and to indicate light sources by creating highlights on objects. They function aesthetically much the same way as textures by adding interest and variety, but can also communicate certain feelings. They do this by mimicking the way light falls on an angled surface: bright in one area, gradually fading into darkness. For example, a gradient background that steps from dark red to light orange mimics a sunset. Dark blue to white or light blue gives the impression of a clear sky fading to the horizon. Lighter areas of the screen attract the eye and project out onto the screen; darker areas fade toward the background.

Gradients are popular with computer designers both for their ability to impart a realistic or natural feel and because they are relatively easy to do. On paper or canvas artists create gradations by carefully blending strokes and changing the color subtly along the way. Using a computer drawing program, on the other hand, a start and an end color are assigned, the program is told how quickly to make the transition happen and in how many steps. The artist can define the direction of the gradient as well. The computer calculates the gradient pattern and applies it to the selected object (see Figure 7-11).

Emotional Associations

As the phrase "He saw red!" indicates, color can have profound effects on individuals. The effects can be physiological, psychological, and/or cultural. Certain colors also can take on meaning for an individual, as in the case of someone who becomes superstitiously addicted to a color for its "good luck" properties.

Table 7-3 cites a few of the general associations human beings have to specific colors as suggested by the *Symbol Sourcebook*. Use these

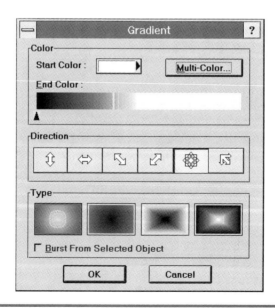

Figure 7-11   Graphics programs make it simple to add gradients

COLOR	POSITIVE	NEGATIVE
Red	warmth, life, joy, happiness, love, passion, blood, liberty, revolution, patriotism	wounds, pain, spilled blood, burning, death, war, anarchy, Satan, danger
Orange	warmth, fire and flames, marriage, hospitality, benevolence, pride	malevolence, Satan
Yellow	sun, light, illumination, intuition, intellect, supreme wisdom, highest values	treachery, cowardice, debauchery, malevolence, impure love, illness
Green	nature, fertility, sympathy, prosperity, hope, life, immortality, youth	death, lividness, envy, jealousy, disgrace, moral degradation, madness
Blue	sky, day, calm waters, thinking, religious feeling, devotion, innocence, truth, justice	night, stormy seas, doubt, discouragement, cold, sadness
Purple	power, spirituality, royalty, love of truth, loyalty, empire, patience, humility, nostalgia	sublimation, martyrdom, mourning, regret, penitence, resignation, humility
Brown	earth, soil, fertility, nature	poverty, barrenness
Gold	sun, majesty, riches, honesty, wisdom, honor, first place, (bronze, third place)	idolatry, greed, commercialism
Silver	purity, chastity, test of truth, moon, second place or platinum (premium before gold)	(none listed)
White	day, innocence, purity, perfection, rectitude, wisdom, truth	spectral, ghostly, cold, blank, void, winter
Gray	maturity, discretion, humility, penitence, renunciation, retrospection	neutrality, egoism, depression, inertia, indifference, grief, old age, penitence
Black	mighty, dignified, sophisticated, regal, fertility, determined, night, solemnity	morbidity, nothingness, despair, night, evil, sin, death, sickness, negation

Table 7-3   Negative and Positive Associations of Colors
(Source: Symbol Sourcebook)

guidelines when adding color to your presentation, but realize the actual connotations will vary according to the audience, the message, the environment and the context in which they are used.

The Power of Sound

In a demonstration/lecture setting, an individual can learn from only two senses: vision and hearing. It might seem odd, then, that sound is so often overlooked as a means of conveying information and stimulating emotion.

The reason for the apparent oversight is simple: the subconsciously influential powers of sound are not well understood by most people. Further, it can be difficult to quantify the contribution of sound to a presentation. Forget any ambivalence or uncertainty—sound works. It can tell an audience what they should *feel*, what they should *think*, and even what they should *do*.

Sound can find its way into a multimedia presentation as narration, music, or sound effects. Any one, or an artful combination of two or all three, can not only add interest and entertainment value to your presentation, it can increase your audience's retention rate and foster a professional image for you and your organization.

Sound design has long been an important part of the filmmaking process. In many respects, it is equally important to multimedia presentations. In essence, your goal will be to design your use of sound, including music and atmosphere, or ambiance, so that it integrates with the visual elements of the presentation. In film, the sound designer conceives an overall strategy for adding sound—when to use music, sound effects, narration—as well as the emotions and style it should convey. The sound designer then works with the director to integrate the sound design with the director's overriding vision.

That may sound like a complex process reserved for professionals and artists, but there is no reason you cannot accomplish the same objective. By learning how sound works to influence an audience and understanding a few of the operative principles, you can enhance both the style and the impact of your presentation.

You might begin your aesthetic investigation of the power of sound by tuning in to a kids' cartoon show on television. Watch it for a minute, then turn the sound all the way down. Notice how your perception of the events changes. You instantly become more critical, less involved. The cartoon may even lose most or all of its humor. You can try the same experiment with any show. Watch the final sequence of *Casablanca* with and without sound. Notice that when you lose the music score, you also lose emotion.

The same principle is at work in multimedia presentations. Adding sound may seem to be extraneous at first, with little relevance to communicating the message. It is not until music, effects, and narration have been added that suddenly the production comes alive.

Sound becomes all the more important as presentations use motion elements in the design. By virtue of the laws of nature and experience people expect things in motion to make sounds. If a ball bounces across the screen, the audience will expect it to make a sound as it hits. A plane flying into view will obviously require the appropriate sounds, but what about a corporate logo that flies on screen?

Narration

Good narration is no different than good speechmaking. Most of the same rules apply to both. The one major difference is that the narration is prerecorded, meaning you may not be able to make adjustments at the last minute. This difference calls for good design and planning.

Some tips to help you use narration effectively:

- Decide which words best describe your key points.

- Know which points you wish to have precedence.

- Decide what tone the narrator should assume.

- Select words that have a clear relationship to the image.

- Determine how the narrator's voice will complement other audio elements.

Writing

Good narration is a matter of imparting the maximum meaning in the fewest words. Effective narration does not dictate, it persuades. It also does not merely provide facts to be memorized, it sparks ideas and paints pictures. Good narration describes more than what can be seen or inferred from the visuals. It draws attention to important points that might be overlooked. It complements the visuals by bringing to the audience more information than what they can see (see Figure 7-12).

Narration must also be coordinated carefully with the images it accompanies. Figure 7-13 demonstrates how altering the timing of the

Figure 7-12 Narration adds extra information to a still image
(courtesy of the Hypermedia Group, Emeryville, CA)

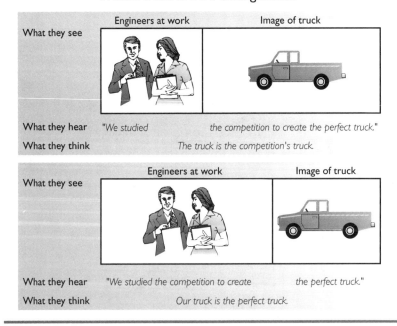

Figure 7-13 Timing of narration and images produces varying
audience perceptions

images changes the way the audience thinks about the subject even when the line of narration remains the same.

tip

Once the script for narration has been written, have somebody read it aloud. It should flow easily and give the reader no difficulty. It also should be immediately clear and comprehensible.

Pacing

Most narration works best when partitioned into complete thoughts (or chunks) that can be addressed individually and kept separate for editing purposes. It is also necessary that the narrator speak slower than in conversational speech. Narration that is too fast for audience comprehension will cause an audience to tune out.

To avoid narration that sounds "written," a certain level of improvisation during the recording session can help. Have the narrator improvise or paraphrase some of the narration, concentrating on the message rather than the quality of the delivery or the literal meaning of the words. Plenty of room must be given for pauses and use of inflection. Failure to vary the pace of the narration will cause wordsandideastoruntogether and the audience will soon lose concentration.

Tone

The feeling or attitude that you impart with narration is as important as the information itself. While tonal quality is highly subjective, the ultimate goal should be to find a "voice" for the presentation. That is, the narrator should seem to be speaking for and through the text, graphics, pictures, and other elements in the presentation. The audience needs to forget about the presence of the narrator. Keep in mind that the moment the narration calls attention to itself your subjective link with the audience will be severed. Also, be certain that the tone of the narration does not clash with the message. A serious subject can be trivialized by narration that seems detached, slightly sarcastic or, in the worst case, whimsical. Have you ever seen a television news cast in which the reporter fails to make a clear tonal transition from a light and humorous story to a serious and tragic one?

Vocal quality

The same words with different voices will have varying effects on your audience. You should never fail to give thorough attention to the type of voice you choose. If you are to be the narrator, be critical of your own vocal qualities—are you the right voice for the job?

Begin your selection process by imagining the different possibilities, then try to zero in on one or two qualities that you feel are most important to conveying the tone and message. You might want to design a test like this to help you decide what you are looking (or listening) for:

Choose the best narrator for each of the following situations:

1. *A multimedia advertisement for a new brand of peanut butter*
   a) Alistair Cooke
   b) An anonymous adult female
   c) Robin Leach

2. *A multimedia sales guide to expensive resort properties*
   a) Alistair Cooke
   b) An anonymous adult female
   c) Robin Leach

3. *A multimedia history of the industrial revolution*
   a) Alistair Cooke
   b) An anonymous adult female
   c) Robin Leach

Then ask yourself why you made your selections. How would the multimedia presentation change if narrated by the other two?

Sound Effects

Sound effects are the punctuation marks in the paragraphs of experience: the sounds that something makes; the background noise that is always there; the unexpected audible intrusions; the aural clues to what we are seeing. You are familiar with the use of sound effects in films and on television. In the golden days of radio sound effects were the "visuals" for radio drama. When they are good, they enhance every element of a scene. When they are bad, they can turn a serious drama into a farce.

In multimedia presentations sound effects play no less an important role, though they are used far more sparingly. They act like aural bookmarks, calling attention to important facts and ideas. A few of the benefits:

> Adding depth and richness to a scene (ocean waves, singing birds, humming machinery)

> Adding humor (a slide whistle as an animated bar graph depicts rising earnings)

> Reinforcing visuals (a ringing register, a cheering crowd)

> Motivating a change of view or scene (we hear the crack of a bat and the image changes to a scoreboard flashing, "It's a home run!")

> Telling a story without words (an empty room; sound of door opening; rapid footsteps; sounds of a struggle; a scream...the sales meeting begins)

tip

Skim through the sound effects samples on the included CD-ROM. Just hearing certain sounds will spark ideas. Make notes on your first reaction and possible applications for the sound in your next presentation.

Music

Music is one of the few things on this planet that requires no translation. It is present in some form in every culture. Its history predates most civilizations. It can communicate feelings, trigger emotions, call forth memories.

Before "talkies" added dialog to movies, music was the language of the medium. When sound tracks were added, music remained. Why? Because filmmakers already understood what musicians have known throughout history—music can convey messages that cannot otherwise be spoken. William Congreve said it has charms to soothe a savage breast.

With luck you will never face an audience of savages—or perhaps you do every day—but should the need arise, music can come to the rescue.

Since well before the advent of digital recording and multimedia, music has played a major role in presentations, particularly those that had a high profile and demanded maximum impact. Yet, until digital audio, MIDI, and clip media came along, scoring a presentation remained nearly as complex and costly minute for minute as scoring a film. Today, presenters and producers have access to much of the same evocative power that filmmakers have used for more than a century, but the costs are coming down even as ease of use goes up.

Granted, business presentations are not supposed to compete with the score of *Dr. Zhivago,* but the use of music to enhance communications is an opportunity that should not be missed. Aesthetically, the influence of music is fairly obvious. Instinct alone should tell you why you want to use music, but when you do, there are a few rules and guidelines to heed.

Musical Components

Music can assume either an active or passive voice. An *active voice* makes direct correlation to the visual elements and timing sequences of the presentation. For example, a collage of images can be timed to change on selected beats of the music. A *passive voice*, by contrast, plays in the background without a direct connection to the individual events in the presentation, such as mood music playing under a narration.

As with narration, discussed earlier, one of the keys to the effective use of music is selecting the right tone. In this context tone does not refer to the audible properties of the music itself but to the overall impression, attitude, or feelings it conveys. Music establishes tone by the consistent use of musical style, choice of instruments, pacing, and tempo.

While a book about presentations is no place for a lesson in music theory, it should be noted that different keys convey different emotions and should be selected with care. A major key, for example, will be associated with good feelings, mainstream ideas, and normal situations. Minor keys tend to suggest sadness and moodiness, but can go further to suggest fear, discord, and conflict. A change in key signifies change in mood or situation. Different instruments and groups of instruments, too, have different connotations. Banjo music, for example, creates a casual, homespun feeling; a brass band conveys energy and exuberance; a

Dixieland band suggest revelry and good times; orchestral anthems signify pomp and circumstance.

The way the music is put together, or edited, will have an effect on the emotions it conveys. Rapid cuts and strong tempos, for example, create a feeling of driving energy and urgency. Slow dissolves and flowing music give a more considered and leisurely feel.

As you begin to explore the aesthetic possibilities of music for multimedia presentation keep the following warnings in mind. They are based on the three most common mistakes presenters make when adding music:

▶ Don't use music as a crutch for weak or insufficient information.

▶ Don't use music to induce cheap or false emotions.

▶ Don't use music when it contradicts the feeling of the visuals or narration.

Mixing and Editing

On average, our eyes blink once every four or five seconds—thousands of times a day. So accustomed are we to this momentary loss of visual information, we hardly ever notice. The same is not true for hearing—since this sense never blinks, we are more likely to notice an imperfection in a soundtrack than a missing frame of video. Our daily experience is made up of one sound environment transitioning to another (going through a door, closing a window). The audio experience in a presentation must be smooth to remain believable. It is the *mixing* of sound that makes this happen.

Effective mixing can do some or all of the following:

▶ Blend several sounds over time

▶ Hide unwanted pops and clicks

▶ Keep sound levels in balance

▶ Process sound to create effects or moods

▶ Balance narration with background sounds and music

Some of the techniques of sound editing parallel those of video editing (see Chapter 10). Table 7-4 lists a few pointers to keep in mind.

The Magic of Motion

We live in a dynamic world—a world of motion. Even if you sit in a room where nothing ever moves, you still have the ability to change your perspective by moving your chair, your head, or even just your eyes. You are, in a sense, the cinematographer (and editor) of what you see.

Notice how during a conversation, you occasionally glance away from the person you are speaking with, perhaps in response to some movement on the periphery of your vision. You might react to a noise and quickly shift your view in that direction. These conscious and subconscious moves are your edits. You are making choices about how you will view the scene. The visual framing affects what you see, what associations you make, with whom you sympathize, and what point of view you take. These experiences are the perceptual basis for the language of film.

Multimedia presentation does not demand that you become a movie director, cinematographer, or editor. But understanding how film works

TECHNIQUE	PURPOSE
Abrupt sound cuts	Disruptive and jarring, especially when connecting two sounds of differing volumes. Draws attention to important transitions.
Fading or cross-fading	Smooth transitions between two tracks minimizes audience distraction and keeps the presentation flowing.
Overlaid third "ambient" track (background noise)	Useful for masking harsh sound cuts or imperfect transitions. Also, helps eliminate the sanitary "studio" quality of narration.
Third track with literal audio source	Sound appears to be coming from source on screen, e.g., radio or musical instrument. Attracts attention, smoothes transitions.

Table 7-4   Sound Editing Tips

to communicate on a conscious and subconscious level will give you the knowledge you need to use motion to influence and persuade your audience.

Just where does the power of motion originate? How and why does it have such a profound effect on people? Experts postulate that human beings, like most animals, evolved instincts for detecting motion as skills for hunting and self-preservation. When we see something move we are instantly alerted to its presence so that we may determine if it is a friend, a potential food source, a threat, or some combination of the three.

As a business presenter you can exploit this natural instinct. As long as the motion is not gratuitous, more action on the screen generally means more information being conveyed and more excitement being generated. Adding motion to a presentation can do the following:

- Hold the interest and attention of your audience

- Communicate abstract concepts and ideas

- Stimulate emotional responses

- Create subconscious associations

- Overcome the modern tendency toward short attention spans

Cinematic Techniques

Videotape, film, and digital video give you the ability to capture motion, store it, and replay it later. In effect, it makes reality portable. But the "reality" you are conveying is your *subjective reality*. The material you choose to show, the selections you capture and edit, introduce a subjective point of view. That is what a filmmaker does from the first frame to the last, even in a documentary.

It is important to remember that the impact and harmony of the motion sequence is derived not from one element or an arrangement of elements but from the effect of the interactions of all the aspects of composition, motivation, rhythm, and pacing. This coordination and blending of elements is sometimes referred to as the visual *choreography*. As with the technical elements of dance, choreography of video and animation in a multimedia presentation consists of developing a conceptual idea of the whole performance, and then instructing individual elements where to move, how to move, and when to move.

Composition

Composition in motion is analogous to composition for static graphics. It describes the way elements are arranged within a frame or on the screen. Good composition is more than just making pretty pictures. It is the organizing force that visually dramatizes and projects the ideas being expressed. The difference between static composition of a still or a screen and dynamic composition of an animated or video sequence is that objects and backgrounds relate to the space and each other over time (see Chapter 1). Motion causes their relationship to be constantly in flux, constantly changing.

As with static graphics, how the eye tracks and "reads" the scene is extremely important to the conscious and subconscious impact. The following are some ways you can use motion to control the focus of your audience:

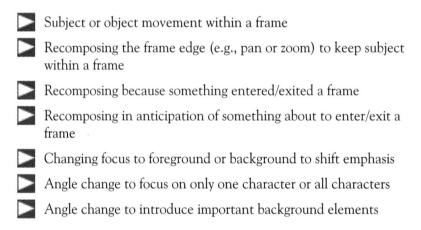

▶ Subject or object movement within a frame

▶ Recomposing the frame edge (e.g., pan or zoom) to keep subject within a frame

▶ Recomposing because something entered/exited a frame

▶ Recomposing in anticipation of something about to enter/exit a frame

▶ Changing focus to foreground or background to shift emphasis

▶ Angle change to focus on only one character or all characters

▶ Angle change to introduce important background elements

The use of camera *angles* and *directions* goes far beyond the literal function of exhibiting or revealing something on the screen. How you use angles and directions will communicate certain meanings or trigger subjective perceptions in your audience.

Ascending compositions, particularly from lower left to upper right, communicate a sense of growth or progress. Conversely, descending compositions from upper left to lower right indicate decline or shrinkage. As a simple example imagine a video sequence in which the camera is focused in tight on an important report. If someone picks the report up off the desk and lifts it into the air, the audience is given a visual clue that the news is good. If the report is thrown down onto the desk, the audience senses the negative direction. The very same principles apply to animating charts or graphs. Figure 7-14 depicts the directions of movement that an audience will subconsciously perceive as growth and progress.

Composition between segments is sometimes referred to as the hidden language of film because it deals with the way different screen objects or different scenes interact. It is one of the most powerful subconscious aspects of motion and can be used to influence perceptions, expectations, and judgments. By carefully sequencing and juxtaposing motion elements you can, to some degree, control your audience's emotional and intellectual response to the event.

For example: You open your sequence in an office. A panning shot to the right past identical cubicles of frustrated office workers leads the viewer's eye to the right side of the screen. Even before anything happens, the movement creates a sense of anticipation or expectation. This sets the viewer up to see a smiling man or woman stepping out of the elevator holding a package of office productivity software. The left to right movement of the camera meets the right to left movement of the character, establishing focus and signifying the importance of the event that is about to occur.

In addition, entire stories can be told between two consecutive shots by building a cause and effect relationship: the first shot is the "cause,"

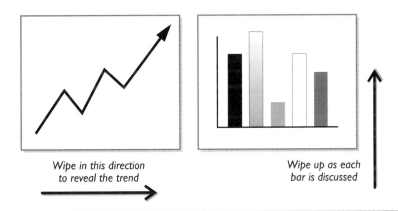

Wipe in this direction
to reveal the trend

Wipe up as each
bar is discussed

Figure 7-14   Use directions to influence audience perception of the message in animated charts and graphs

the second shot the "effect." Even though the interim story is missing, nothing is lost. In the above example, if the final shot were to show the office again, but this time everyone is working happily and productively, it is not necessary to show what happened after the person stepped off the elevator. The effect of the action has been clearly demonstrated.

tip

The point of concentration at the end of one shot should lead the eye to the point of concentration at the beginning of the next shot.

Motivation

When an audience sees action on the screen, they consciously and subconsciously begin developing opinions about the motivation behind the movement. If, for example, you depict an executive walking down an assembly line, the audience may interpret that as the executive's interest in staying in touch with the work force. A slow gait could show thoroughness; a fast walk might indicate a get-it-done attitude. The same techniques work for inanimate objects. If you have your logo speeding into view, it may connote a sense of urgency and vitality. What impression might your audience get if, instead, you let the logo grow from a tiny speck, getting larger and larger until it completely dominates the screen? Not to suggest that it is all about speed—the motivation suggested by a change in angle, a shift of perspective, or a transition sequence can be as evocative or more so than the subject itself in conveying your message.

As you begin to work with the power of motion keep the following basic principles in mind:

▶ The viewers' eyes will automatically lock onto moving objects, even when initially presented on the periphery of the screen.

▶ Moving objects will dominate inanimate objects within a frame.

▶ Camera movement, subject movement, and transitions all coax the eye in the direction they move.

▶ Movement can occur with the subject (walking, running, jumping, etc.), the camera (tracking, dollying, panning, tilting, zooming in, zooming out, etc.), or both.

Multimedia in Action

The color images in this insert are reproductions of the images accompanying the case studies in each chapter. Of course, these images can't convey how the actual presentations combined color, shape, sound, and movement to create a total impact greater than the sum of the parts. Nevertheless, the images offer a glimpse of the variety of rich and textured environments that can be created using multimedia technology for presentation.

GTE
Chapter 1

GTE advanced its communications capabilities with the creation of an integrated information system that combines multimedia sales presentation capabilities with a multimedia training program for field personnel. Video, sound, graphics, and real-time cost calculations solidify the link between the company and its customers by providing current product information and highly customized services.

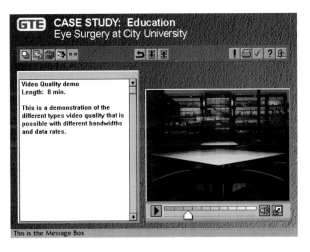

UNION BANK OF SWITZERLAND
Chapter 2

UBS of North America invested in a multimedia presentation system to add value to a presentation delivered before the company executives in Zurich. While a series of colorful animations riveted the attention of the audience, hidden interactive buttons allowed the presenter to bypass inappropriate sections, jump to areas of greatest interest, and drill down into deeper levels of detail on the fly.

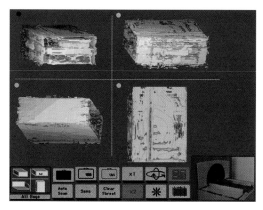

INVISION TECHNOLOGIES
Chapter 3

InVision Technologies demonstrates the functions of its explosives detection system to airports and other potential clients worldwide by transporting a 30-pound multimedia computer instead of its 3-ton security device. The multimedia sales presentation simulates an emergency bomb threat on video before switching to a graphical representation of the inner workings of the detector.

XIRCOM
Chapter 5

Xircom networks with its customers and salespeople using a modular multimedia system that adapts to sales calls, training sessions, and marketing presentations. The modules are tied together thematically by an animated character who guides the user through the presentations and demonstrates procedures for using Xircom products. The modular structure of the system allows salespeople to tailor presentations specifically to their prospects.

USX CORP.
Chapter 6

USX Corp. helps its employees solidify their retirement plans with a user-friendly multimedia program that employs brightly-colored graphics, sounds, and animation to the difficult journey through the land of 401(k) plans, social security, and pensions. Dubbed EASY (for Employee Access System for You), the program links directly to the company mainframe and does all the necessary calculations, allowing employees to experiment with different scenarios using real-time information.

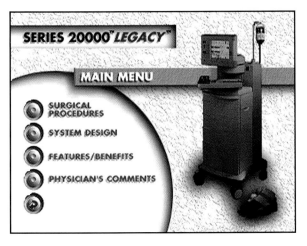

ALCON LABORATORIES
Chapter 7

Alcon Labs targets its customer base with surgical precision using a CD-I-based information kiosk that contains an hour of full-motion video demonstrating the operation of the company's medical equipment. The graphic design of the presentation as well as the kiosk housing is coordinated with the colors, style, and design of the products themselves.

TURNER BROADCASTING SYSTEM
Chapter 8

TBS went the distance with an elaborate multimedia presentation promoting multi-million dollar sponsorships of the Goodwill Games sporting event. In addition to video, an original soundtrack, and narration by actor James Earl Jones, the presentation featured 3-D models of the sports venues in St. Petersburg, Russia, which depict the potential sponsor's logo as it might be seen during the actual event. A hidden button allowed the sales team to load a customized presentation for each audience.

LEE APPAREL COMPANY
Chapter 9

Lee Apparel dresses up the boys department of R.H. Macy's in Manhattan with a stacked-cube multimedia display designed to lure kids with vivid cartoon images, pop music, video, and quirky sound effects. The kiosk adds an element of interactivity by inviting the potential young customers to enter their height and weight into the computer, at which point the program tells them the jeans size they should buy.

FIRST DATA
Chapter 10

First Data equips its large, mobile sales force with multimedia laptop computers and an extensive multimedia sales presentation. Using digital video, graphics, sound, and interaction, the program demonstrates the sophisticated information services of the First Data Health Systems group by staging a mock videoconference between two doctors and a patient.

NABISCO FOODS
Chapter 11

Nabisco adds a pinch of spice to its trade show displays with interactive kiosks that not only provide food buyers background on new and existing products, but display and print out flavorful recipes and cooking hints as well. The kiosk collects data directly from the attendee's show registration card, allowing follow-up information to be mailed to the potential customer even before the trade show.

HEALTH NET
Chapter 12

Health Net takes care of its own with an in-house multimedia kiosk that lets employees access up-to-date information about their health plans and make necessary changes on the spot via a link to the company database. In addition, employees can browse a database of primary care providers to learn more about available physicians, including photos, biographical information and even a map to the doctor's office. The presentation is kept light and entertaining by health tips that appear throughout the program.

Still Motion

Some of the dynamism of full motion can be imparted to still images using a motion-controlled camera. The technique, commonly seen in historical documentaries where little or no historical video is available, uses special mounts and motors to move the camera in and out (zooming) or side to side (panning) on the fixed image. The viewers eye, to some extent, interprets the camera movement as motion within the scene. Motion-controlled shooting is a demanding skill and best left to professionals. But you might consider having a segment produced and digitized for addition to your multimedia presentation.

Another way to add the sense of motion to still images is with the use of transition. *Cross fading*—the blending of one scene into another—for example, is a technique commonly used in film and video and will give your still transitions a cinematic feel.

Rhythm and Pacing

Visual rhythm is defined as the frequency of shot changes and how often the camera moves. As in music, this creates a tempo that can be used to have a desired effect upon the audience. Regularly spaced transitions at a leisurely pace, for instance, will create a steady, reassuring (or monotonous) rhythm. Staccato, irregular shifts will create a jarring and chaotic rhythm. Your goal should be to match the rhythm both to the content and the feeling you would like to spark in the audience.

Determining the *optimum pace* for any shot is a subjective process, but a good test for pacing is to think about the speed at which your audience will be driving past your "billboard"—if they zip by too fast they might miss the message. If they go too slow, they might have too much time to examine, critique, and possibly reject the message.

The factors that determine the *optimum duration* for a shot or a sequence of shots include not only the content of the message and its complexity, but also the audience's expected tolerance. Your goal, as with the entire presentation, is to make the segment long enough to tell the story, but not so long you lose your audience. If during the process of determining the duration of your motion sequences you have not thrown out some footage *you* really love, it is probably too long.

Aesthetic Sensitivity

As explained at the top of the chapter, no one is a master of all arts, but everyone has the capability of being sensitive to aesthetics and aesthetic influences. Aside from formal training or study, there are simple things you can do to gain a better understanding of the subtextual functions of media. Begin by noticing each time you react emotionally to a movie, a piece of music, a sound, or the narration on a television documentary. Make a few notes about your physical and emotional response. Did you react to the literal content of what you heard or saw, or did you react to something less obvious?

Try to identify the root of your reaction, and then determine if the media was manipulated in such a way as to evoke your reaction. Do you think your response was what the artist had in mind? Or was it contrary to the intention of the work? Try and figure out how they did it, and what techniques were brought to bear.

Ideally, you will have the opportunity to witness other traditional and multimedia presentations from time to time. Employ the same process. Pull back from the subject matter and the media elements for a moment. First identify which media are being used, then try to determine why. Finally, alert yourself to the attitude, mood, or perception the presentation has communicated. Use what you learn as you become more proficient with using multimedia as a tool for persuasion.

The most brilliant colors, spread at random and without design, will give far less pleasure than the simplest outline of a figure.

—Aristotle

PICTURE THIS: WORKING WITH GRAPHICS

Multimedia presentations are inherently visual. Photographs, illustrations, shapes, and colors on the screen combine to create a visual language that speaks to an audience both subtly and directly. The more fluently you speak the language of graphics—the greater your visual eloquence—the more successful you will be at communicating your message.

Graphic Impact

Graphics, as they relate to multimedia production, include just about anything visual that appears on the monitor, except video. Graphical elements work not only individually but together to generate the overall look or feel of a multimedia presentation. Well-implemented, harmonious graphics give the presentation an identity and cohesiveness that informs, instructs, and delights an audience.

Here are just a few of the most obvious uses of graphics:

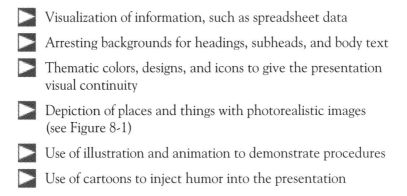

Visualization of information, such as spreadsheet data

Arresting backgrounds for headings, subheads, and body text

Thematic colors, designs, and icons to give the presentation visual continuity

Depiction of places and things with photorealistic images (see Figure 8-1)

Use of illustration and animation to demonstrate procedures

Use of cartoons to inject humor into the presentation

One of the great advantages of digital production is the capability to create, import, and edit everything from photos to animations to pie charts. Indeed, many of the presentation software packages on the market today are referred to as presentation *graphics* packages. Their initial function was and still is to give business users convenient and flexible access to the communicative powers of graphics. If you are already working with one of these packages, then you are familiar with creating charts and graphs, importing clip art, using templates, and maybe doing some painting and drawing.

Graphics can either be created inside your multimedia presentation software or in an external application program designed specifically for the task. Provided that the platform, file format, and storage requirements are met, you can import virtually any type of visuals into your presentation as you build it. You might add a simple organizational chart, for example, by drawing it with the paint and draw tools available in presentation software such as PowerPoint, Persuasion, or Freelance. Or, you could use the more powerful and precise tools in an illustration

Figure 8-1   Scanned photographs of people, places, and things
add power and dimension to a presentation
(courtesy of Medior, Inc., San Mateo, CA)

program such as CorelDraw. Or, you might find the exact design you need in a clip art collection on CD-ROM. You even have the option of scanning an organizational chart from a book or document and importing it into your application as a graphics file.

The method you choose for any graphics operation depends on the available materials, your skill level, time factors, the graphics software tools, and the specific look and the desired effect you are after.

In this chapter, you will explore your creative options as well as learn about the graphics handling realities of personal computers. Whether you have any formal training in art or graphic design, with a little effort and practice you can incorporate elements into your presentations to make specific points and create a pleasing visual experience for your audience. A basic understanding of computer graphics will empower you to make creative decisions and conform to the production values you have set for your multimedia presentation.

Graphics Realities

Before you make any decisions about hardware or software, you should know a little about how graphics are handled in the computer. Typically, a computer and graphics software create, manipulate, and store graphics as one of two types: bitmap (also called "raster") or vector. With *bitmap graphics*, the software looks at whatever is on the screen and writes a location and color value for each dot that makes up the image. The dots are represented in the computer as bits (the use of multiple bits to store color information is discussed in the next section). *Vector graphics* work in an entirely different way than bitmapping, using mathematical formulas instead of defining each bit of an image.

note

The dots of color in a bitmap or vector graphic are frequently referred to as pixels. A pixel is a short way of saying picture element, and it refers to the smallest identifiable piece of information in a bitmap image. Pixel is also used when referring to the dots on a monitor screen.

If you cut, paste, or transfer a bitmap image, all the computer code for each of those thousands of pixels goes with it. Imagine a drawing on a piece of paper. Regardless of bending it, tearing it, folding it, or mutilating it, all the picture information is still there until you physically throw a piece away. The paper continues to occupy the same volume of space.

The fact that bitmap graphics are exact maps of every bit contained in the graphic becomes apparent when you examine the effect on file size. A photo-realistic image that fills a standard computer screen will require approximately 922K of storage—nearly one megabyte per picture. In multimedia production, storage space is often a critical issue. Later in this chapter you will see exactly why bitmap file sizes are so large and find out what you can do to keep these graphics files down to a manageable size.

With vector graphics, on the other hand, the software program defines the outlines or boundaries of the image shape as well as information about what fills it, then stores the coordinates and spatial relationships in a mathematical formula. For that reason, vector graphics work best with images that are primarily made up of lines and shapes as opposed

Technical Primer # 1

Bitmap vs. Vector

To get an idea of how vector graphics work, picture a red rectangle in the center of the screen. The coordinates stored in memory or in the file would read something like "RT:100,100,540,380,12". RT de-fines the vector object type as a rec-tangle. The first coordinate or corner of the rectangle is $x=100$ and $y=100$, or 100 pixels from the left of the screen and 100 pixels down from the top of the screen. The second x-y coordinate pair defines the opposite corner of the rectangle as 540 pixels from the left and 380 pixels from the top. Finally, the last number refers to a color. In this case, color number 12 is red. Fortunately for the nonpro-gramming user, the computer does the math and keeps track of changes.

The vector file for the red rec-tangle would require only a few bytes; even with all the accompanying in-formation from a drawing software program, the total would be no more than about 8K. A bitmap file for the same rectangle would require 440 pixels horizontally by 280 pixels ver-tically, or 123,200 pixels. Using 8-bit color (the effect of colors on file size will be discussed in the next section), the file size would be approximately 123K.

to photorealistic images that rely on form, tonal gradations, shadow, and texture. The primary advantage of vector graphics is the relatively small amount of data and the resultant small memory and file size needed to describe the image

As shown in Figure 8-2, bitmaps and vector graphics can coexist on the same screen. The photorealistic bitmap graphic of the currency in the image was imported into the presentation. The programs' paint and draw tools were used to create the "funny money" vector graphic.

Resolving Resolution Issues

In addition to understanding how computers draw, paint, and store graphics, it is important to understand the concept of resolution. Reso-lution is a term that has become muddled in its reference to computer graphics because of its dual meanings. Someone looking at a graphic and asking "What resolution is that image?" is actually asking two questions. Resolution refers to the number of colors that can be displayed for each

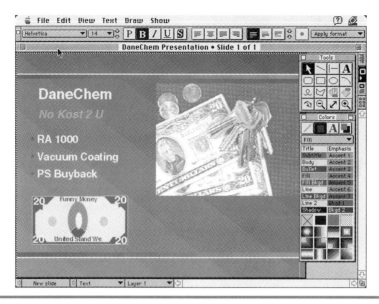

Figure 8-2   Presentation images frequently combine bitmap and vector graphics

pixel (color resolution); it also refers to the total number of pixels that make up the image (video resolution).

Color Resolution

The *color resolution* of an image refers to the total number of colors possible for individual pixels in the image, and is defined mathematically by the number of digital bits the computer uses to record the color of each pixel. To keep terminology straight, most graphics professionals refer to the color resolution as *bit depth*. One-bit graphics, for example, allow the pixel to be represented in only two possible color states—black or white.

Adding another bit doubles the range of possible colors to four, and so on. When the computer uses eight bits to represent one pixel or dot in the image, the color resolution is referred to as 8-bit. Eight-bit color is in fact one bit (two options) to the eighth power, or 2^8, which yields 256 possible colors. For reference, the math works out this way: 8-bit color represents 2^8 or 256 possible colors; 16-bit color represents 2^{16} or 65,536 possible colors; and 24-bit color translates to 2^{24} or

16,777,216 (16.7 million) colors. Table 8-1 demonstrates the relationship between bit depth, the number of possible colors, and the effects on file size. It also shows the video resolution standards that apply to the various bit depths at a 640×480 video resolution. Video resolution is discussed in the next section.

The specific colors that are available for reproducing the image are referred to as the *color palette*. Image reproduction generally improves as the number of colors in the total palette increases. Most people, for example, will see a visible improvement in quality when looking at a photorealistic image drawn from a 16 million-color palette versus a 65,000-color palette, because the 16 million-color palette has more colors available to fill in the subtle gradations in hue.

Nevertheless, image quality using a 256-color palette can be quite good for color graphics if a *custom color palette* is defined for each unique image. A custom color palette optimizes the colors that are used to reproduce an image so that none are wasted. Most color graphics cards will allow you to create custom palettes. If custom palettes are used, fading to some neutral color such as black between images in an onscreen presentation is necessary to avoid the problem of color shift. *Color shift* occurs when the old color palette is applied to the new image, or vice versa. The effect can be a startling disruption of the colors on the screen.

Some graphic boards in the PC and Mac environments are capable of 24-bit color resolution delivering more than 16.7 million colors. These are sometimes referred to as *true color* boards. True color can be a

BIT DEPTH	POSSIBLE COLORS	VIDEO STANDARD (AT 640×480)	APPROXIMATE FILE SIZE*
1 bit	2 colors	Monochrome	37K
2 bit	4 colors	CGA	75K
4 bit	16 colors	VGA	153K
8 bit	256 colors	XGA, SVGA	307K
16 bit	65,536 colors	XGA, SVGA	614K
24 bit	16.7 million colors	True color	922K

*File sizes are theoretical and given for comparison only. Actual file sizes will vary with image and graphics software.

Table 8-1   Sample File Sizes at Various Bit Depths

Technical Primer #2

32-Bit Graphics

You may run across a few graphics programs that implement ultra high-quality 32-bit graphics file formats. These files are most frequently in one of two forms. The first is a 32-bit CMYK file (CMYK stands for the Cyan, Magenta, Yellow, and blacK inks used in four-color process printing). While this format is commonly used for print advertising, its color system has no advantages for screen-based multimedia, which works with the red, green, and blue values of each pixel.

The second type, file formats such as the TGA (Targa) file format associated with the Truevision series of color graphics boards, reproduces 32-bit color by combining 24-bit color with an 8-bit alpha channel. The alpha channel holds an 8-bit color or grayscale image that can be used to modify the 24-bit color values. For example, suppose you have a photograph of your company's product and you want to transparently superimpose it over an 8-bit beige stucco background. In order to experiment with various effects, you keep the 8-bit alpha channel image of the stucco intact while experimenting with the 24-bit image on top of it You see them combined on the screen, but the computer keeps the information separate for each. Alpha channel techniques are most commonly used by artists and designers using Adobe Photoshop and similar image editing programs.

significant enhancement to a multimedia presentation. If your computer processor is powerful enough and your budget allows it, a 24-bit board is a good investment in high-quality production values. Note in Table 8-1, however, that true color generates correspondingly large file sizes.

Storing and Processing

The quality level of the graphics will directly affect the size of the file. Generally, the greater the color bit depth, the larger the file size (see Table 8-1 for a comparison of file sizes at different bit depths). The 24-bit images in particular are not only storage intensive but also tax the capabilities of a computer's central processor. An 8-bit image uses one byte of storage for every pixel. A 24-bit image uses three bytes per pixel, so the image file becomes three times larger; switching a 300K graphic from 8 bits to 24 bits increases it to 900K.

If you opt for the high quality of 24-bit true color graphics, keep in mind that your program will run slower and your storage demands will be higher. One compromise chosen by many presenters is to create and import graphics in 24-bit on the development system, then save them as 16-bit or 8-bit images with a custom palette for playback purposes. Using this strategy, the original image retains maximum resolution, a fact that could be important if you plan to use the image at a later date in a more powerful presentation delivery system. Image editing programs such as Adobe Photoshop (see Figure 8-3), Aldus PhotoStyler, and Corel-PHOTOPAINT give you the option of saving at a lower color resolution.

Video Resolution

Video resolution refers to the total number (not colors) of pixels your graphics display board and monitor are capable of displaying. One way to think of resolution is as a matrix of columns and rows. A typical graphics board and monitor conforming to the VGA standard (see the "Graphics Standards" section following) has 640 columns and 480 rows of pixels, or 640×480 resolution (see Table 8-1).

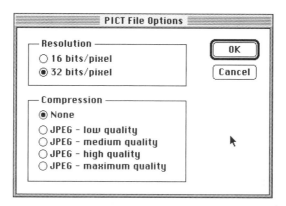

Figure 8-3   Adobe Photoshop lets you save an image at various bit depths

note

The resolution capabilities of the graphics board must be compatible with the resolution of the monitor. A high-resolution image (1024×768), for example, cannot be displayed unless both the graphics card and the monitor support that level.

Graphics Standards

Most graphics boards support multiple video resolutions. Generally, the more expensive the board, the higher the resolution capability. The following standards or display protocols have been developed as guidelines for compatibility and capabilities. They are listed in ascending levels of video resolution quality.

▶ VGA: The Video Graphics Array, or VGA, standard used with IBM PC and compatibles describes a graphics capability of 640×480 video resolution with 4-bit (16 colors) color. This is a basic-level standard that conforms with MPC Level 1 standard but not MPC Level 2 standard (see Chapter 3 for more about MPC standards).

Technical Primer #3

Color Resolution and Video Resolution

At this point, understanding the relationship between color resolution and video resolution is important. A 640×480 display contains 307,200 pixels. If the color resolution is set to 8-bit (remember, an 8-bit image uses one byte of storage for every pixel), then the representation of the screen will require 307,200 bytes. Increasing the color resolution to 24-bit, as shown earlier, triples the number of bytes to 921,600 or approximately 922K. Maintaining the 24-bit color depth and increasing the video resolution to 1024×768, the size of the image grows to 1024×768×3, or 2,359,296 bytes (2.4MB).

XGA: The Extended Graphics Adapter standard developed by IBM in 1990 provides for 8-bit color up to 1024×768, or 16-bit color at 640×480 pixels.

XGA-2: A 1992 IBM standard that provides for 8-bit color at 1024×768 and 16-bit color at 640×480.

SVGA: Super VGA, or SVGA, was developed to answer the demand for graphics reproduction with more colors and higher screen resolution. Generally, it refers to cards capable of multiple resolutions, including 640×480 resolution at 4-bit, 8-bit, or 16-bit color; 800×600 resolution at 4-bit, 8-bit, or 16-bit color; and 1024×768 resolution at 4-bit or 8-bit color. SVGA conforms to the MPC Level 2 requirement for 16-bit color at 640×480.

note

While the 800×600 and 1024×768 screen resolutions of SVGA look very good on a desktop monitor, the benefits may be lost during projection of a multimedia presentation. Increasing the number of potential colors, on the other hand, can be highly desirable. A good course of action with multimedia is to buy an SVGA board and set it to run at 640×480 with 16-bit color.

Workstation Resolutions: Many graphics cards are available that will exceed the screen resolution of even SVGA. 1280×1024 resolution is fairly common for engineering workstations where highly detailed Computer Aided Design or CAD drawings are being created. These resolutions are also ordinarily unnecessary for multimedia presentations.

QuickDraw: The Macintosh has until recently only had one video resolution display protocol, known as QuickDraw. A new version, QuickDraw GX, is due soon. Meanwhile, most Macintosh machines can support 24-bit color at 640×480.

Graphics Acquisition

With a little training, even individuals with no serious background in art and design can learn to create effective graphics and imagery for multimedia presentations. The beauty of computer illustration is that if you tell a computer you want a circle it will let you draw a perfect circle far more precise than the freehand sketch of any artist—lines will be straighter and colors purer, too (see Figure 8-4). Of course, artistic skill is more than a matter of round circles and straight lines.

If you do not have the inclination to create graphics yourself, or you are simply a graphics basket case, you have a number of effective alternatives that can yield good-to-outstanding results. You should begin the graphics creation process by carefully evaluating your own skill level and asking yourself the following questions:

1. Do all the cars on the road seem to be incorrectly stopping at *green* lights? (Color perception varies widely between individuals, and "color blindness" is relatively common.)

2. Do you have in your living room any paintings of Elvis in Day-Glo colors on black velvet? (Artistic taste is a function of personality, training, and cultural orientation.)

3. Do you find IRS forms graphically appealing? (Visual design is both an objective and subjective skill.)

4. Do you have any items of apparel that cause others to confuse you with a crossing guard or circus performer? (Color sense is both instinctual and learned.)

5. Do all of the individuals in your photographs seem to be missing limbs or foreheads? (Almost anyone can be trained to understand basic composition skills.)

If you answered yes to any of the above, then you might want to consider a course in graphic design, photography, or art appreciation. Dozens of good books on design are available as well. A recommended option is to invest in a quality collection of clip art and/or the assistance of a talented freelance graphic artist (see the following sections, "Digital Clip Art" and "Outside Help").

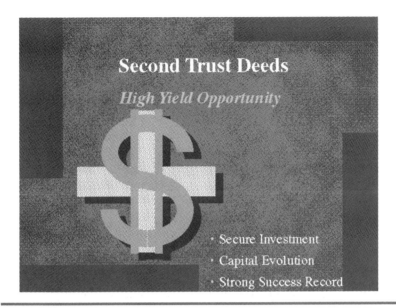

Figure 8-4   With a little practice using the drawing tools in presentation software programs, almost any user can create basic graphics, such as the dollar sign depicted here

Digital Clip Art

Before the advent of computer graphics, graphic artists would use clip art books to add illustrations and design elements to an advertisement or other work whenever they wanted to eliminate the need to draw or paint from scratch. The books are called "clip art" because the artist would clip out the image and paste it onto a page layout for printing.

Digital clip art has made the process not only easier, but far more flexible. Images are stored on CD-ROM or floppy disk and can be electronically clipped and pasted into any program that has graphics handling capabilities. Electronic clip art collections are available on virtually any subject. Some have literally thousands of still images, ranging from photographs to cartoon characters to simple black and white icons. Many presentation and other graphics software packages come with clip art collections included. A sampling of clip art is included on the CD-ROM that comes with this book (see Figure 8-5).

To use clip art, you have to understand the basics of file formats (see the "File Formats" section later in this chapter). The presentation or graphics software you are using must be compatible with the file format of the clip art collection. Some collections have proprietary file extensions, others conform to one of a dozen or more standard extensions. In cases where your software does not handle a particular file format, you have the option of using a file translation program. These software utilities translate the file from its native format into one your system is capable of recognizing.

On the Macintosh, the most common clip art format is called PICT. In Windows, BMP is the most common format; EPS clip art is also common but is primarily designed for print applications. (For further discussion, see the "File Formats and File Handling" section of this chapter.) Clip art can be stored at a variety of bit depths: 8-bit, 16-bit, and 24-bit. Some clip art packages store the same image in two or more resolutions, allowing the user to select the file size and quality level that best suits the application.

Figure 8-5   Sample clip art from the CD-ROM that accompanies this book

warning

When using clip art, read the rights agreement carefully. It is unlawful to use, copy, or reproduce copyrighted material without permission. The agreement for use may be limited in number of reproductions or types of projects, or it may require a royalty payment. It may also require referencing or crediting the clip art vendor or the artist.

Outside Help

Deciding whether to attempt to create graphics yourself or to hire assistance is a simple matter of balancing your time, tools, and talent against the desired production values of the multimedia presentation. In-house graphics help is, of course, the first course of action. Failing that, outside help is available for every aspect of graphics production, from simple illustration to complete design services. Make certain you see sample work that corresponds to the style and quality you are seeking. The best artist or designer is one who will take the time to understand the entire project, including goals and budget limitations.

Ask that artwork be delivered electronically in a file format you are capable of handling. Even if the artist is working in traditional materials, you might want the artist to be responsible for scanning the work first, thus saving you the trouble later. Make sure your resolution and bit depth requirements are clearly understood. If you are hiring a design firm, ask for references and check them carefully. Get specific commitments and have a thorough understanding of what will be done if the work is not satisfactory as delivered. For more on using outside talent, see Chapter 12.

Graphics Utilities

Every presentation software package on the market includes at least basic painting and drawing tools. You will be able to draw shapes, then size and color them. Most presentation packages will also automatically create charts and graphs. A few offer impressive 3-D and animation effects as well. Some of the packages have extensive toolsets, with such sophisticated effects as color gradients and texturing. If you are working with a media integration package that is focused more on importing and arranging media for presentation than on creating it from scratch, you will likely have to depend on external graphics software and clip art libraries for your images. The graphics tools for high-end authoring systems also cover the spectrum. Some have only the most basic graphics capability, while others have extensive toolsets.

For do-it-yourself graphics, the tools you use will depend on your skill level and the quality you wish to achieve. What follows are brief descriptions of the most common graphics tools. Remember, not all packages will give you all the tools. You will also find most of these tools and techniques demonstrated in the tutorials at the end of the chapter.

Painting and Illustration

In general, the main difference between painting and illustration programs is that *painting programs* create and work with bitmapped graphics, while *illustration programs* create vector graphics (recall the discussion on bitmapped and vector graphics at the beginning of this chapter). Painting programs use software brushes, pens, and other artistic tools to generate computer "painted" images. Texture, shadow, and brushstroke are typically emphasized in the painted image. A painting program creates images with soft edges and painterly effects. Gradients, gradually changing of hues from one pixel to the next, are typical of the effects in a paint program. Figure 8-6 shows some of the painting tools available in Fractal Design Painter.

Vector-based drawing or illustration programs also employ a variety of graphics tools, but in general they focus on creating lines and shapes. An illustration program is ideal for creating graphic design elements in a presentation. Drawing tools let you use the computer's precision to create lines and shapes that are mathematically perfect. Examples include CorelDRAW (see the tutorial at the end of this chapter), Computer Support Arts & Letters Express, and Adobe Illustrator (shown in Figure 8-7). These programs and others like them do everything from text treatments to complex perspective drawings. They support gradients, pattern fills, and custom colors. Being vector-based, software in this category creates color graphics that can be stored using less disk space than painting (bitmap) programs.

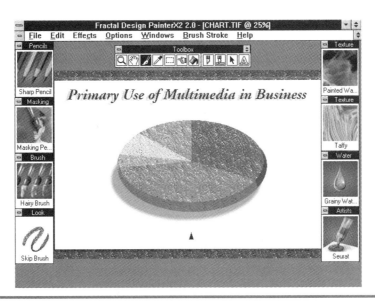

Figure 8-6   Paint programs such as Fractal Design Painter, shown here, give you a variety of artist's tools for creating and altering images

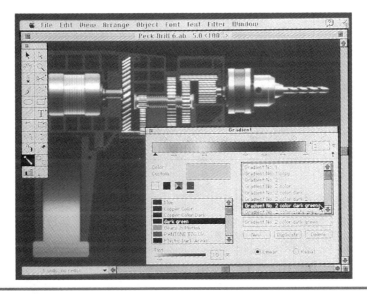

Figure 8-7   Drawing programs such as Adobe Illustrator, shown here, are excellent for creating complex drawings

Image Editing

Specialized painting programs that are designed primarily to edit scanned photographs and images are referred to as *image editing programs.* The tools in image editing packages allow you to correct technical deficiencies in a photograph or other image. The images can also be altered or enhanced with a variety of artistic effects. The clear software leader for Windows and Macintosh in this category is Adobe Photoshop (see the tutorial at the end of this chapter), although Aldus PhotoStyler, CorelPHOTO-PAINT (shown in Figure 8-8), and Picture Publisher are popular as well.

The programs can adjust such technical elements as brightness, contrast, and color balance. Most programs have tools for blurring or sharpening an image as well as moving or removing elements. The programs use filters and alpha channel techniques to, in effect, hide certain characteristics and overwrite the visible image with the changes; using this technique, the illusion of a shadow and other effects not

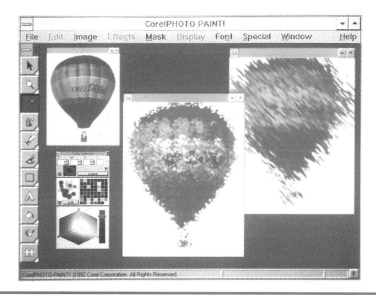

Figure 8-8 Using CorelPHOTO-PAINT, this photo was given an "impressionist" and a "pointillist" painting look with the aid of filters

contained in the original can be created. Alpha channels can also be used to apply texture, color, and transparency effects.

Some image editing software programs allow you to add "plug-in" extensions that broaden creative capabilities. Generally, the plug-ins are for advanced users, but if you have a good working knowledge of the main program, the extensions can prove useful and powerful. One example of an extension is Kai's Power Tools for Photoshop from HSC Software. Figure 8-9 shows a cover of *Business Week* magazine that was created using Kai's Power Tools and Photoshop. The image was later used in a multimedia presentation.

Text Handling

Text is one of the most important graphical elements on the screen. Its proper use both aesthetically and technically can make or break a presentation (see Figure 8-10). It is the element you will likely use most often and most directly.

Figure 8-9   Program "plug-ins" such as Kai's Power Tools for Photoshop extend the paint and photo manipulation capabilities of the host software

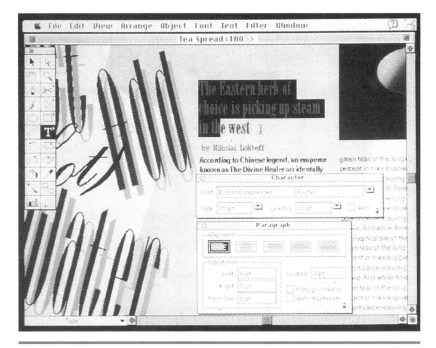

Figure 8-10 The text shown here was manipulated in Adobe
Illustrator to create custom looks

Just about any graphics program will have some provision for creating
and editing text. Most will let you choose from either the fonts that come
with the program, the fonts you have on your system, or both. For all
practical purposes, fonts come in two forms: PostScript and TrueType.
PostScript uses a series of files at specific sizes that describe the font at
the particular size. TrueType uses a single font file that contains detailed
geometric information on how the letter forms should look at a wide
range of sizes. The font that is used in creating the presentation will have
to be loaded on the delivery station as well, otherwise the delivery system
will default to a font that may destroy the look of the type on screen.
Painting programs can convert fonts to bitmaps as they are placed into
an image, eliminating the threat of missing fonts during the presentation.

Font styles generally can be set in all the ways familiar from standard
printing: bold, underlined, or italicized. Many programs also offer outlined or shadowed text capabilities.

note

The font shadow option available through the Font menu on most programs is not the same as a drop shadow. Drop shadows are an artistic treatment of text accomplished by offsetting a darker duplicate of the text in the background (usually at least two pixels below and to the right or left of the foreground text).

When using text and graphics together, the outline of the text can sometimes look jagged—the square pixels give the edge a stair-step appearance. This unwanted effect, referred to as *aliasing,* occurs particularly when text is set on a highly contrasting color. The effect can be partially corrected using anti-aliasing filters. Few presentation programs have anti-aliasing filters. To smooth the text, exporting the image to an image editing program may be necessary.

The selection of fonts and the design issues surrounding type are primarily matters of aesthetics (see Chapter 7). Generally, although hundreds of fonts are available, you will probably only need a half dozen or so at your disposal to create the look you want. Once you find the fonts that work, stick with them and do not get *too* creative, or you may end up with a screen that looks like a ransom note.

Animation

Animation packages and capabilities range widely, and selecting the right tool can be confusing. But what is not difficult to understand is the power that animation can adds to a multimedia presentation. Even the simplest animated chart or screen transition adds a new dimension of impact to a presentation.

Animation tools can be found in presentation slideshow programs, such as Astound or Compel. Animation tools are also integral to some media integration packages such as Cinemation or GoldDisk Animation Works. High-end authoring programs, such as Director, frequently include animation capability. Add to that programs designed to do nothing but create professional-level animations, such as Autodesk Animator and a dozen or more 3-D modeling programs with animation capabilities.

Case Study:

Company: Turner Broadcasting System, Atlanta, Georgia

Business: Television news and entertainment

Objectives: Design a traveling multimedia presentation to help convince corporate prospects to sign up as sponsors for the 1994 Goodwill Games

Turner Broadcasting System (TBS) understood from previous experience that convincing U.S. corporations to invest millions of dollars in sponsorships for its 1994 Goodwill Games sporting event requires master salesmanship. While the sales effort for the Games had been successful in the past, the company faced the new realities of a dissolved Soviet Union and post–Cold War economic uncertainties. They needed to give potential sponsors a preview of the St. Petersburg venues and generate enthusiasm for the event based on both the commercial and philanthropic benefits.

According to Craig Apatov, vice president and general manager of corporate marketing for TBS, a powerful—and costly—multimedia production, complete with full-screen, full-motion video and 3-D graphics, was determined to be the best way to let potential sponsors see and experience some of what they could expect from TBS's integrated marketing approach: a blending of signage, promotion, licensing, commerce, and hospitality.

In the presentation's introduction, the Macintosh computer running Macromedia Director software cues a videodisc-based video sequence in which actor James Earl Jones narrates a history of the Goodwill Games and provides a video tour of historic St. Petersburg. Other pieces of the presentation include broadcast and advertising schedules, as well as marketing opportunities, promotion support, and sports venues.

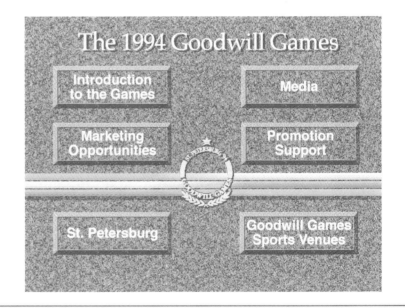

Turner Broadcasting System

Once the mood has been set with images of striving athletes and a worldwide audience, the sales team goes for the gold with the presentation's most persuasive and imaginative sequence. A menu selection opens the screen to a 3-D map of St. Petersburg (created using Dynaperspective and Stratavision 3-D modeling and rendering tools) depicting the four main sporting venues. Clicking on any of the models, the target sponsors are "transported" into a realistic rendering of the facility, where they see their corporate logos prominently displayed in various locations—above the entrance, on the scoreboard, around the ice rink, on banners, wrapped around kiosks, even on a blimp hovering above the running track. As they view the stadiums and arenas from the perspective of the spectators and television audience, the sponsors also hear the "roar of the crowd," created using clip media sound effects and Macromedia's SoundEdit Pro editing software.

The custom-venue effect was developed using Adobe Photoshop to size and shape the corporate logos to fit onto the 3-D models. Separate screens were designed for each set of logos and brands belonging to each of the potential corporate sponsors. In advance of the presentation, the presenter opens a hidden menu in the program and selects from the list of potential sponsors. The program automatically loads the appropriate image files.

According to Apatov, the presentation allowed TBS to present a much clearer idea of what the Goodwill Games represent and what the event means in terms of exposure and marketing opportunities. It has been shown to more than 200 companies around the world. Most importantly, however, TBS has been able to secure more than a dozen sponsorships and has "eclipsed the level of sponsorship achieved in 1990," says Apatov.

Most of the tools that are accessible to the nonprofessional animator use *path-based* animation techniques; briefly, these programs let you define a route for an object by dragging it across the screen. The program records the movement as a series of frames and plays the frames back rapidly to simulate motion. More sophisticated programs let you define *key frames* at certain stages of the animation; the computer calculates the necessary shape and number of images to fill in between. Animations are typically controlled either by using a timeline that determines the total duration and how fast or slow the animation will play, or by setting the frame rate. With the latter, the speed and duration are dependent on the processing speed of the computer. A few packages, such as Cinemation, also provide cel-based animation techniques, similar to the way professional animators use layers of transparencies, or *cels,* to add changes in an object's position. See the tutorial later in this chapter on using Cinemation.

Presentation software animations for bullet points, slide builds, and motion transitions such as wipes and dissolves can in many cases be added to presentations with the click of a button. The number of transitions and animation options vary. Most packages will also allow you to play an animation that has been created independent of the program.

3-D Modeling

Three-dimensional (3-D) software refers to software that uses three-dimensional vectors to define the boundaries or shapes of objects. The 3-D geometry used is called the *model.* This model can be assigned light sources, and the surfaces of the objects within it can take on textures. The specular and reflective properties of the model's surfaces can be further specified by the user in some programs. Once all of the visual properties are assigned to the model, it can be *rendered* (calculated and drawn) from any viewpoint. *Ray tracing* occurs when the 3-D program is sophisticated enough to reflect imagery from one object off of another. (See the tutorial at the end of this chapter.) A 3-D animation is created by stringing together a sequence of images or frames. Each frame of the animation contains a different 3-D rendering, usually with a shift in the point of view, change in the motion in the model itself, or both.

3-D typography programs, such as Pixar Typestry, Specular International LogoMotion, and Crystal Graphics Flying Fonts, allow the creation of models from type fonts and other 2-D sources. The models are rendered with a variety of texture and lighting effects. They can also be animated. The finished animation is played as all or part of a scene or slide in the presentation.

Virtus Walkthrough is an example of a 3-D program that can be used to create both simple and complex 3-D environments. A unique feature of Virtus Walkthrough, as the name implies, is its ability to navigate through the models on the fly. Programs such as this can be used in a presentation to demonstrate spatial relationships of objects from the point of view of the audience.

File Formats and File Handling

The PICT standard file format for the Macintosh environment made file handling on the platform a snap from the beginning by setting a platform standard. It is one of the primary reasons that the Mac became and remains a favorite of graphic artists. Most any Mac-based graphics program will export and import files in PICT, making compatibility almost a nonissue.

For the PC, the situation has been nearly the opposite. Since graphics first appeared on the platform, dozens of formats have been created, some compatible, some not. With the introduction of the Windows interface for the DOS operating system, file handling has improved significantly. Windows is standardized on the BMP file format, and most external graphics programs have adopted BMP as a main format or format option. Nevertheless, you may want to use graphics from other sources that were not originally intended for Windows or multimedia applications. To a lesser extent, the same may also be true for the PICT Macintosh format.

File translators have already been mentioned, in the "Digital Clip Art" section of this chapter, as one file-handling option. Third-party products, such as DeBabelizer for the Macintosh, will convert a broad range of files from one format to another with relatively frequent success (see Figure 8-11). Converters are also built into many painting, illustration, image editing and presentation software products. Whether you will have the exact conversion utility you need for your situation will depend both on the source file and the destination program. Because DeBabelizer supports both Macintosh and PC graphics file formats, you can use it to translate files from a format suitable for one platform to a file format suitable for the other, but only if you are running a Macintosh in the first place (see the "Cross-Platform Options" section following).

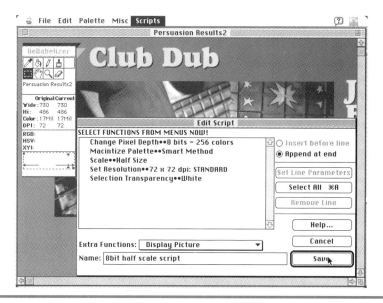

Figure 8-11   DeBabelizer uses a script-editing dialog box to set translation parameters

Image Management Software

In order to manage the huge numbers of graphic files that you will generate when creating numerous multimedia presentations, you should consider using an *image management software* program. These programs let you create catalogs with thumbnail representations of your graphics. You can specify a variety of search criteria to help you find the image at a later date. Aldus Fetch for the Macintosh, UImage Pals for Windows, and Kudo Image Browser (see Figure 8-12) for both platforms are examples of programs that help you keep track of your images.

File Formats

As mentioned earlier, every graphic file conforms to a format specified by the software developer. In order to maximize the ability to use a graphic created in one program within another program, certain interchange formats or universal file formats have been established. Some formats are the result of agreements by ISO or ANSI standards committees. Others are commercial standards that have evolved because

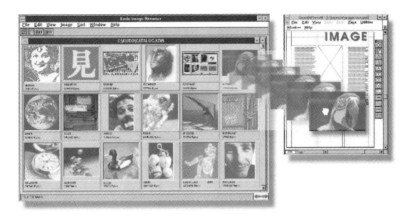

Figure 8-12   Screen shot of the Kudo Image Browser user interface

of the popularity or originality of a particular software or hardware device. Table 8-2 lists the most popular and common file formats along with their file extension designations.

As mentioned in Chapter 5, a key factor when chosing your authoring software is the program's ability to import the file formats you will be using most often. Format-handling capabilities vary from package to package. Most multimedia-capable authoring programs give you a good choice of formats (see Figure 8-13).

Cross-Platform Options

Working *cross-platform* (sharing files between Windows and Macintosh) can be confusing and frustrating if you do not plan carefully for both the physical transfer of the files as well as the translation of the file formats.

The physical transfer of the files can be accomplished on floppy disk, removable hard disk, across a network, or using modem communications. Because 3.5-inch floppy disks are now nearly universal, this is a

FORMAT NAME	FILE EXTENSION	TYPE OF FILE
Windows Bitmap	.BMP	Bitmap
Drawing Exchange File	.DXF	Vector
Encapsulated PostScript	.EPS, EPSF	Vector
GIF	.GIF	Bitmap
GEM file	.IMG	Bitmap
Initial Graphics Exchange	.IGS, IGES	Vector
JPEG	.JPG, JPEG	Compressed bitmap
Lotus Pic	.PIC	Vector
PCX	.PCX	Bitmap
PhotoCD	.PCD	Bitmap
PICT/PICT2	.PCT, PICT	Bitmap/vector
Raster Image File Format	.RIF	Bitmap
RLE	.RLE	Compressed bitmap
Targa	.TGA	Bitmap
Tagged Image File Format	.TIF, TIFF	Bitmap
MS Word Metafile	.WMF	Bitmap
Word Perfect Graphics	.WPG	Bitmap

Table 8-2   File Formats

convenient method if your files are small. A floppy disk holds up to 1.44MB of information per disk. You can exchange files written on a Macintosh with a DOS or Windows machine using programs such as Apple File Exchange, Access PC, Mac in DOS, or DOS Mounter. The floppy disk is formatted in one machine with the format of the other. Then files are copied onto that disk. The disk can now be read normally on the other platform, though the file formats may have to be converted before the application program can import them.

Some removable hard disk media formatted for DOS can be written to or read on the Macintosh as well. For example, SyDOS is a version of the popular SyQuest disk cartridge (see Chapter 11) and can be read on a Macintosh as long as the proper system extension is installed.

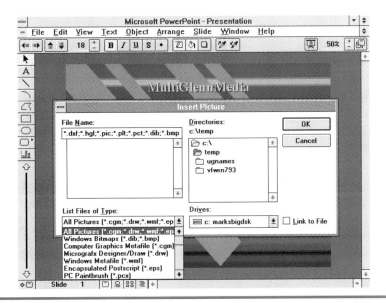

Figure 8-13 Most programs use a dialog box to let you see the file import capabilities as you scan your directories for images

Many network file servers based on the Ethernet protocol serve both Mac and PC users. With Novell Netware and Netware for the Macintosh, the file server can be accessed by cross-platform users on the network.

Modem communications conform to standard protocols, which can be used by both Macintosh and PC users either through direct connections or by passing electronic mail with file enclosures on bulletin board services, such as CompuServe. As the digital so-called information superhighway and the Global Information Interchange (GII) come on line around the world, transferring large graphics and multimedia files to most anywhere will soon be relatively easy.

Output Options

When dealing with multimedia presentations, most graphics are designed to be displayed onscreen or projected in some way. You may also need to output some or all of your images to some other form, either as backup or as handouts. Transparencies and slides are the most obvious backup media. Finding either an overhead projector or slide projector in most presentation venues is relatively easy. Should the multimedia delivery system fail (not an entirely unlikely possibility), a backup set of visuals could save your presentation.

Slides

Film recorders, either in-house or at a service bureau, can convert your computer graphics images into 35mm slides. These computer peripheral devices can give you high-quality resolution. Output will vary according to the quality of the system and the software compatibility. Most film recorders work with major slideshow software applications as well as with standard graphics file formats. Keep in mind that the only multimedia elements that transfer to the slides are the static graphics. If you want to simulate animation, a series of slides can give your audience the general idea, but that's about it.

note

The ratio of height to width of a computer screen is 3:4, while the ratio for slides is 2:3. Most presentation packages allow you to create your images with slide dimensions, but if you do not intend slides as your final output, you probably will use the full-screen dimensions. If images are to be transferred to 35mm film, they will have to be resized, or some of the visual information will be lost.

Transparencies and Color Handouts

Most color printers will output both transparencies and color handouts. Special transparency films are available for thermal wax and ink-jet printers. You will achieve a resolution of about 300 dots per inch (dpi) for most. Some color printers have a resolution as high as 360, 400, or even 1,200 dpi. The cost per page for color output usually ranges from 25 cents to several dollars per page, depending on the process used. For large quantities, you may want to use a service bureau to create color separations, which your printer can then use to create large quantities of copies. For medium-sized jobs, you might even consider printing the first color copy on your color printer and then having duplicates printed on a color copier. A black-and-white laser printer will often be adequate for printing handouts. All of the programs mentioned in this chapter can print to standard laser printers and color copiers.

Tutorials

The following tutorials are intended to give you a feel for how the products in each category operate. They are not meant to teach you how to use specific products; only the product manual or, perhaps, a book dedicated specifically to the product can do that. Nor are the tutorials meant to suggest the best way to perform the operations depicted.

Drawing: CorelDRAW

CorelDRAW is a popular Windows application that can be used for creating illustrations to include in your presentations. It was chosen here because the tools and techniques it employs are top-of-the-line examples of those you will find in most drawing programs. CorelDRAW is only one of the tools that comes with the Corel product; others include the CorelPHOTO-PAINT image-editing module, the CorelSHOW presentation tool, and CorelMOVE animation software.

In this scenario you are going to create a single graphic image for a presentation about the Far West Welding Supplies company that is hypothetically intended as part of a larger multimedia production. Remember, many of the graphical elements you create in CorelDRAW could also be done directly in your presentation software program, but CorelDRAW is far more powerful when it comes to tools, effects, and controls.

Step One: Enter text

After opening CorelDraw, pull down the Layout menu. Choose Page Setup. Click on Paper Color and select black. Click OK.

Open the Effects menu and select the Extrude Roll-up. Open the Text menu and select the Text Roll-up.

Now, click on the Text tool (the capital "A") in the Tool palette. Place the cursor roughly where you want the type to appear and click once. From the Text Roll-up dialog box, pick a font from the list. In this case, a sans serif font called "Bahamas Heavy" is used, but any font will do. Set the font size to 100 points by entering **100** in the size window or by scrolling up with the arrow keys provided. Move the cursor to the bottom of the screen and click on a yellow color in the horizontal palette. Now, click on Apply in the Text Roll-up.

Place the cursor roughly where you want the name to appear and click. Type **Far West Welding Supplies**. In the illustration below, the company name is entered near the top, but some headroom has been left so the title can be rotated.

If your font is too large, highlight the text on the screen, select the Text tool again, and change the size until it roughly matches the illustration.

Now reselect the Text tool, and choose a smaller font size of 36. Place the cursor near the bottom, click, and type "**Far West Is Far Best**".

Step Two: Select the text

Reselect the company name by double-clicking somewhere near the center of the text. Notice that when you do this, the text is selected with *rotation handles* (small arrows on each corner).

Step Three: Skew the text

Now select the up and down rotation arrows on the middle-right portion of the image. By dragging this handle up, the text will be skewed so that each character retains its upright orientation. Slant the text upward about 15 degrees:

Step Four: Draw a circle

To create the company logo, select the Ellipse tool (the oval shape) in the Tool palette. Place the cursor at the left below the title text. Hold down the CTRL key (the CTRL key allows you to draw perfect circles) and drag the cursor to draw a circle beneath the lower-left corner of the company name, as shown here:

Step Five: Draw two lines

Set the width and color for the lines by clicking and holding on the Stroke tool (it looks like a fountain pen) in the Tool palette. Select a medium-wide line. Set line color from the same tool by clicking on the color wheel (it looks like a spoked wheel). Select yellow for both lines.

Now, click on the Freehand tool (it looks like a pencil) and draw two lines (the color is still set to yellow) so that they start outside the circle and blend into it—click once where you want the line to begin and again where you want the line to end. Next, use the Selection tool (the arrow at the top of the Tool palette) and click once somewhere off the drawing so that nothing is selected. It should now look something like this:

Step Six: Draw a contrasting zigzag

To change the color for the next set of lines, click and hold the Stroke tool (fountain pen) again. This time, select the color wheel at the lower-left of the Pull-across palette. Select black as the color. Notice that you can also choose black by clicking on the black box in the palette. Keep the line width the same as you selected for the yellow lines.

Click on the Freehand tool (pencil) and use it to draw a black zigzag pattern in the center of the yellow circle so that it looks roughly like the zigzag shown here:

To do so, click once at the edge of the circle to start the line. Click (and release) once more where the line is to turn. Click again to draw the angle line. Repeat for the third leg, which should end at the opposite side of the circle. (Note: do not click and *drag* or you will draw a freehand curve.) You have completed the company logo.

Step Seven : Duplicate the logo

Now that you have one version of your logo, you can duplicate it easily by first clicking on the Selection tool (arrow at the top of the Tool palette). Place the arrow at the upper right of the logo, click and drag a box completely around it. Release, and all elements will be selected. Pull down the Arrange menu and select the Group option.

Now pull down the Edit menu and select the Duplicate option to create a copy of the logo. Perform the duplication again to create a third copy. Click on each copy one at a time and drag them into a horizontal arrangement. To make sure they are aligned, select them all by using the Section tool. Choose the Arrange menu's Align option and choose Vertically Centered by clicking the selection box. It should now look like this:

Your graphic is complete. Save your file if you wish.

Image Editing: Adobe Photoshop

Adobe Photoshop is used widely on both the Windows and Macintosh platforms. Not only can it be used to scan photographs and slides with your scanner, you can use it as a painting package for impressive text treatments and artistic effects.

In this tutorial, we will modify a scanned photo of a church that Brad's Painting Company is bidding on. The changes to the photo will show what the building could look like with a new color. To follow this tutorial, select an image from the CD-ROM included with this book or from some other source available to you.

If you wish to use a scanned photo for this tutorial, begin by following the instructions that came with your scanner for scanning a new image. If you use a scanner such as the Microtek ScanMaker, you can choose the Adobe Photoshop File menu Acquire option and control the scanner directly from within Photoshop. Other scanners may require you to scan the file and save it in a separate program before you can begin.

Step One: Crop the photo

Once you have your image on screen, begin by selecting the Cropping tool (a rectangle with a slash through it, at the top left of the Tool palette). The Cropping tool has handle points on each corner of the image, and by clicking directly on these control points and dragging them, you can adjust the size of the crop box before you actually crop the image. Once the crop box is the way you want it, move the cursor to the inside of the crop box, where the cursor changes its appearance to a scissors icon. Click once and the image is cropped—everything outside the marquee (dotted rectangle) is deleted. In the example, the photo has been cropped already to show the church and a bit of foliage on each side:

Step Two: Adjust levels

When you look at the image immediately after scanning, it may appear to have lost detail in the shadows. If so, open the Image menu and select the Adjust Levels option. The dialog box appears:

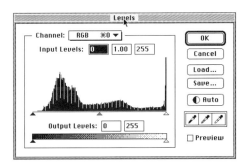

You might see that the combined RGB (Red, Green, Blue) output levels appear to be bunched together in places; if so, click the Auto button. This usually yields excellent results, and the RGB levels now appear more evenly distributed. You may want to adjust the midpoint for the midtones, highlights, and shadows manually as well. Once it looks good, accept the new settings by clicking OK.

Step Three: Select the area to change

Now, select the area to be changed, using the Magic Wand tool (icon of a wand with a sparkler on the end) from the Tool palette. The Magic Wand selects all pixels of similar color. Click on the area to be changed, in this case, the outside walls of the building. A dotted line appears to outline the area.

To adjust the Magic Wand's similarity settings, double-click on the tool, and increase the number to select a larger area or decrease it to select less. The Rectangle and Oval Marquee tools, as well as the Lasso tool, may be used in combination with the Magic Wand to select just the area you want. To add areas to the current selection, hold down the SHIFT key. To remove areas from the current selection, hold down either the Command key (⌘) if you are using the Mac, or the CTRL key if you are using the PC. While the key is pressed, use the additional selection tool.

Step Four: Choose a new color

To replace the old one, click on the foreground color selector (the topmost of the two overlapping squares in the Tool palette), and the Color Picker appears:

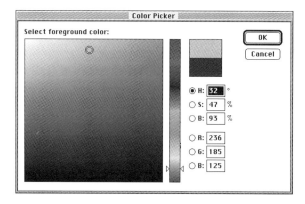

The slider bar picks a particular hue in the spectrum. The large rectangular area allows you to adjust the saturation and luminance of the hue. A tan color has been selected for the church example. Select OK.

Step Five: Fill with the New Color

If the selected area were to be filled now with the tan color, it would lose its photorealism and would instead look like a photo with a tan blob in the middle. To avoid that effect, pull down the Edit menu and click on the Fill option. The dialog box appears and gives you options on how you want to use the foreground color you have picked:

Fill

Blending

Opacity: 100 %   Mode: Color ▾

OK

Cancel

The mode or characteristic you want to change is Color. This action leaves the saturation and luminance values intact, so that the details of shadow and texture are not lost. In this case, Opacity is left at the default of 100%. Select OK.

Step Six: Hide edges

To get a good look at the details in your image, you want to hide the "marching ants" that surround the selected area. Pull down the Select menu and choose the Hide Edges option. Click and the ants disappear.

Step Seven: Add text

Now you are ready to add some text to the presentation image. Choose the Text tool (capital "T") from the Tool palette. Pick the spot where you want the text to appear, and click once on the image. Enter the text, **Brad's Painting**, and set the font, point size, and other text characteristics in the dialog box. Make sure the Anti-aliased box is selected, as this ensures a soft edge between the text and the background imagery. Click OK. When the text appears on the image, move the cursor into the text itself. It changes from a text insertion tool to an arrow pointer. Click and drag the text to the desired position:

Step Eight: Copy the text and fill the drop shadow

With the text still selected, copy the text to the Clipboard by pulling down the Edit menu and clicking on Copy.

Now change the foreground color of the text to black by clicking on the color selection icon (overlapping squares) in the Tool palette (refer back to Step Four). Select black and click OK. Select the Fill option from the Edit menu again, but this time choose the Normal mode at 100% opacity. Click OK. Black text appears in place of the tan text:

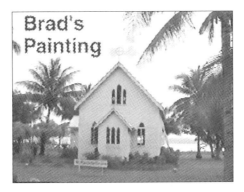

Step Nine: Paste the foreground text

The drop shadow effect is completed by pasting the tan text from the Clipboard on top of the black text. To do so, use the Paste command from the Edit menu. The tan text appears on top of the black. Now use the cursor arrow keys to move the tan text up two pixels and to the left two pixels. Do this by pressing the up arrow twice and the left arrow twice. The black drop shadow is revealed underneath. The text should appear similar to this:

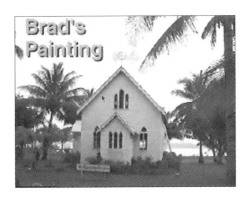

Your image editing is complete. Save the file if you wish.

3-D Modeling: Adobe Dimensions

Adobe Dimensions for the Macintosh is just one of dozens of 3-D modeling programs. Unlike other graphics software, 3-D modeling software tends to vary quite a bit from package to package. Dimensions makes a good example here because of the way in which it effortlessly accepts text and graphics from Adobe Illustrator files and allows you to transform them into professional-quality 3-D renderings. This ease-of-use factor is welcome if you have only a little time to invest in three-dimensional effects. This tutorial demonstrates the steps for creating 3-D text.

Step One: Import the 2-D text and set the view

In this example, the fictitious name "Lorac Industries" has already been typed into Adobe Illustrator using the Helvetica font. The font characters have been modified a bit to give them some extra style, in preparation for export to Dimensions. You can use any Illustrator text file of your choosing for this tutorial.

Open Dimensions. Pull down the File menu and Import the file you wish to use. Once it is in place, pull down the View menu and set the Front View option:

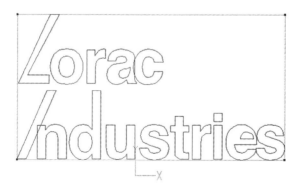

Step Two: Extrude the text

To give your text a depth in the third dimension (the z axis), pull down the Operations menu and click on the Extrude option. With Dimensions, you can work in points or simply express depth as a percentage of height. In the dialog box, select 15% with no bevel. Click OK. Later, as you get more adventurous, you will want to experiment with the Dimensions Bevel Library.

Step Three: Define surface properties

Now you are ready to decide on the surface properties of the text. Pull down the Appearance menu and click on the Surface Properties option to get this dialog box:

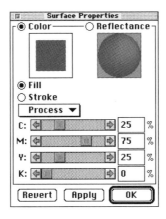

In the dialog box, set the Color properties for the object fill colors. In this case, use the Process color mode, which allows you to define percentages of each color. Set the Cyan (C) value to 25%, Magenta (M) to 75%, and Yellow (Y) to 25%. For this exercise, leave blacK (K) at 0%. The sphere in the upper-right corner of the Surface Properties dialog box will show the general shading characteristics of the color you have defined.

Step Four: Rotate the object

Select the Trackball icon (a ball with a curved arrow around it) at the bottom of the Tool palette. By clicking and dragging the text, you can experiment with different angles until you find the angle of view you are after. Note, holding down the SHIFT key as you click and drag the text object will keep the text aligned horizontally to the point of view. When you have the text positioned as you want it, release the mouse button:

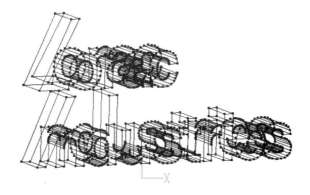

Step Five: Render the scene

You are now ready to render the object. *Rendering* applies the lighting, surface properties, reflection properties, and shading to the geometry from the current point of view. To start rendering, pull down the View menu and click on the Final Render option.

The simple text created here should only take a couple of minutes to render on a Macintosh Quadra.

Once you see your final 3-D text, if you want to make changes, you can choose not to save it and go back to work on the wireframe image. If you like what you see, give it a filename and save it if you wish.

Animation: Vividus Cinemation

Cinemation, which was designed to make it easy for presenters to create interactive presentations with animation, sound, and QuickTime movies, provides three ways to animate. "Click-and-drag" animation allows the user to record simple motion across the screen by clicking on the record button and dragging the object to be animated. "Fill In Motion" automatically moves, scales, crops or rotates an object over any number of frames. "Frame-by-frame" animation can be created using Cinemation's multiple frame ghosting option. Filmstrips provide a frame-by-frame story-board to navigate through movies and to cut and paste frames.

Cinemation users with PowerPoint and Persuasion can use the AutoMotion feature to automatically enhance presentations created previously by importing a file and applying templates to make text and logos slide, swoop, or zoom on the screen in various patterns. This is particularly useful for organizations or departments that maintain librar-ies of presentations.

Cinemation also includes a 24-bit color paint program, the Cine-Player for distributing Cinemation movies, the MovieWindow XCMD for playing and controlling movies from HyperCard, and a collection of clip animation and sounds.

In this tutorial you are presenting the year's sales figures and have decided to add animation to draw attention to the numbers. You will also create a menu screen that allows you to jump to different topics in your presentation.

Step I: Create some text

Open Cinemation to start a new file. The Tool palette and Control panel that you will be using are pictured in Figure 8-14. The program begins with the first frame of the new movie. Click the Text tool (the letter "A") in the Tool palette, then click on the center of the screen. From the Text menu, choose a font, style, color and size for the text you will enter. Type **International Sales $500M**, then deselect the text block by selecting the Arrow tool on the tool palette, and clicking on the blank frame.

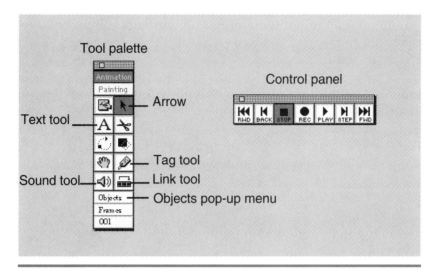

Figure 8-14 Cinemation's Tool palette and control panel

Step 2: Create a background

The default background of a Cinemation movie is white. Select a background color for your movie by opening the Animate menu and choosing Background Color.

Step 3: Add animated elements

From the File menu, select Import. Click on the File Type pop-up menu and choose Movie (Scrapbook format). From the free clip animations included on the CD-ROM disk, select "Earth", then click Import.

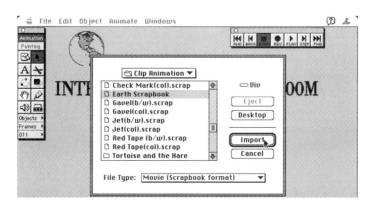

An image of the Earth now appears in the frame of your movie. Select the Earth, go to the Object menu, open the Transparency option, and choose Trimmed. This eliminates the object box surrounding the Earth.

To make the Earth grow in size as if moving toward the viewer, hold the SHIFT key, grab one of the four object handles of the Earth clip, and drag it toward the center of the Earth. Scale the Earth to about 1/4-inch in diameter. Next, click REC on the Control panel, hold down the SHIFT key, grab one of the handles again, and scale the Earth clip back out to a larger size of about two inches in diameter.

Keep an eye on the frame counter in the upper-left corner of the screen as you scale the Earth. It indicates the number of frames recorded. In this case you will want to record 37 frames of movement.

Now that the Earth is scaled, you want it to move from the upper-left corner of the screen, where it appears reduced in size, to the center of the screen, where it will be larger.

Click REW on the Control panel to return to the beginning of this sequence. Select the Earth and drag it to the upper-left corner of your screen. Now click REC on the Control panel, then grab the Earth and drag it slowly to the center of the screen beneath your text. Watch the frame counter and stop on or before frame 37.

To view the finished sequence of the animated Earth, click RWD and then click PLAY on the control panel.

Step 4: Add sound

Return to the first frame of your movie by clicking REW on the Control panel. Click on the Sound tool in the Tool palette to open a simple dialog box that lets you add, import, or record sound.

```
Add Frame Sound

When playing back frame 2,
◉ Start the sound called:          [ Play Now ]
  [ Drum Roll        ▼]
    ☐ Repeat until next sound      [ Import... ]
                                   [ Record... ]
○ Stop sound
                        [ Cancel ]  [  OK  ]
```

Import one of the free sample sounds included on the CD-ROM. Click OK and the sound is added to your movie. The sound will play back automatically when this frame of your movie plays.

Step 5: Identify frames for interactive linking

Cinemation provides a Tag tool (indicated in Figure 8-14) for naming individual frames of your movie and for establishing or altering interactive links between various elements. In order to access the Salesframe interactively, click the Tag tool on the Tool palette, then click on the blank frame and type **International** into the dialog box that opens.

```
Name Frame

Name:  [ International              ]

Frame number: 2     [ Cancel ]  [  OK  ]
```

Click OK. This frame is now identified.

Step 6: Add an interactive menu

Click the arrow tool and select Insert Frames under the Edit menu. Click OK and Cinemation adds a frame to your movie. Next, using the Text tool as before, click on the top right of your screen and type **Sales Summary 1994**. Press the ENTER key and the text cursor opens a new text box immediately below. Type Domestic, press ENTER, and type **International** in this final text box. Now click on the Arrow in the Tool palette.

To make the text item into an interactive button so that during playback of your final movie you can jump directly from this menu to the slide. Click the Link tool on the Tool palette.

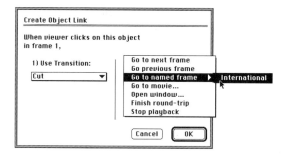

When the Object Link dialog box opens, select one of the frame transitions ("Cut" in this example); then under Link, select Go to named frame and highlight Click OK. You have now created a direct link between the menu item and the animated sequence "International Sales $500M". During playback of your final movie, the animated "International Sales" sequence will play whenever you click on "International" in the menu frame.

To put a pause on this menu frame, begin by deselecting all objects. To do this, open the Objects pop-up menu on the Tool palette and select the No Object option. Now click on the Link tool and select Pause until Object Link. Click OK. During playback, your movie will pause on the menu frame "Sales Summary 1994" until one of the interactive menu items is chosen.

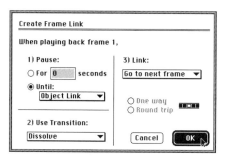

Step 7: View the finished movie

To view your interactive movie, click PLAY on the Control panel. Move the arrow cursor to the line of text. The arrow transforms into a hand indicating that this is an interactive menu item. Click your mouse and your movie branches to the animated sequence "International Sales $500M".

The reason why we have two ears and only one mouth is that we may listen the more and talk the less.
—Zeno of Citium, ca 300 B.C.

THE AURAL DIMENSION: WORKING WITH SOUND

A presentation without sound is like a rainbow in black and white. Breaking the sound barrier is one of the easiest and most rewarding steps in the multimedia process. With digital sound capability you will have access not only to the overt powers of narration, but also to the subliminal influences of effects and music.

Sound Decisions

The one indisputable requirement for any presentation, multimedia or otherwise, is an *audience*. A presentation without an audience is the proverbial sound of one hand clapping. The word "audience" is mentioned here for its obvious kinship to the word "audio." Both are derived from the Latin *audire*, meaning "to hear." The word "auditorium" also stems from the same root.

Though an audience usually comes to the auditorium to see as well as hear a presentation, it would be difficult to overstate the importance of sound. In business communications, in entertainment, and as part of life in general, it shapes our perceptions and is a primary medium of communication.

Playwrights and actors of ancient Greece relied almost entirely on the power of spoken words, vocalization, and music to tell their stories in the large outdoor theaters of the time. Two thousand years later, Shakespeare and his contemporaries still favored voice and music over physical action. Even the silent films of the early 1900s were never silent upon presentation. A picture is worth a thousand words—that may be, but think how much more effective a silent movie is with music to set the mood and heighten emotions. Music-invoked moods, feelings, and emotions are often every bit as important to a business presenter as they were to Charlie Chaplin.

As with films, a presentation has numerous ways to make use of sound, including background music, narration, voice annotation, and sound effects. In this chapter, you will learn about the available tools for exploiting both the conscious and subconscious power of audio. (For a more complete discussion of the aesthetics of music and sound, refer to Chapter 7.)

Do not be intimidated by the blizzard of hardware and software options or by the technical terminology. Computers may not take to sound as well as they do to numbers, text, or graphics; but, in general, once you have your hardware in place and understand the playback and recording issues, you will find audio as straightforward to incorporate into your multimedia presentations as are illustrations and photographs. In most cases, any irritations you experience as you learn to employ computer sound technology will be far outweighed by the benefits.

Sound Advice

Why should you use sound in business presentations? The benefits are clear as a bell:

▶ *Heightened interest and impact:* Music and sound can keep your audience tuned in and focused; under extreme circumstances, they can also keep your audience awake. Supplement an animated revenue chart with the sound of dropping coins, or add a ringing cash register to underscore the net profit figures. If the message is good news, add a fanfare. Bad news, try a foghorn. If the material is boring, jazz it up with a few jazz riffs.

▶ *Better retention:* How many times have you heard a jingle on television or radio and then could not get the tune out of your head? Harness the power that advertisers have profited from for decades to make your business message as unforgettable as "You deserve a break today...." Connect the feeling of the music with the feeling you want the message to convey and, in many cases, that is how the message will be remembered.

▶ *Image enhancement:* We are judged by our tastes in such things as food, clothing, cars, and music. Audio cannot do much for you with the first three, but the right choice of musical enhancements can help you establish in the mind of your audience just about any image you desire for yourself, your organization, or your presentation. Do you wish to be thought of as formal and businesslike, hip and stylish, cool and laid back, classical and refined? Say it with sound.

▶ *Comic relief:* Renowned speakers, clergy, and politicians have always known that moments of great seriousness and profundity are more effective when juxtaposed with moments of humor. It may take nothing more than an offbeat sound effect or a laugh track to put your audience at ease and let them breath deeply for a moment. It's a jungle out there, so why not say so with an authentic Tarzan yell? If sales for a product have come to an abrupt halt, nothing says it better than the sound of screeching tires.

Perceived value: Researchers at the MIT Media Lab had test subjects watch identical television sets with identical pictures. The sound from one set was degraded somewhat. When asked which set had the better *picture quality*, the subjects consistently selected the set with the better sound. The same phenomenon will work for your ideas, products, or services. The higher the quality of your sound elements, the more positive the response to your message.

Getting Started

You should begin your exploration of audio tools and techniques with the knowledge that it is not always as easy as it sounds. Essentially, you will be dealing with three forms of sound: music, narration, and effects. These can come from a variety of sources, and in the case of music, there are two entirely different ways of recording and generating the sounds in your computer. The first challenge you face will be acquiring auditory content.

For music sources, if you are not a musician (or not a *good* musician), you have two basic choices. First, you can hire professional musicians to create an original score (underscore the word "professional"). Relatively few people can appreciate the nuances between an orchestral performance of Mozart by the Boston Symphony versus the Berlin Symphony, but everybody knows bad music when they hear it. Also, your original composition should be recorded on the highest quality tape with good equipment. It will then have to be converted to digital format, a time-consuming and storage-intensive process. Or you could record directly to digital media using the proper hardware, an option not yet frequently used.

Your second option, one that is becoming increasingly easier to exercise, is to use prerecorded music or *clip music*. Extensive CD-ROM collections are available for musical recordings, MIDI (see its definition and discussion later in this chapter), and sound effects, and you will save a fortune compared to commissioning original work. Until recently, not much was available, and every presentation that used clip music sounded painfully similar. Now you can select from a wide range of quality compositions, including classical selections.

tip

The quality of clip music and effects varies dramatically. Some of the clips are obviously recycled from bad recordings or are almost unusable because of shoddy recording or digitizing. Do not buy the cheapest disk—it's cheap for a reason. Look for samplers and free demos, such as those included on the CD-ROM with this book, that will let you hear the quality of the collection you are buying.

Rights and Ownership

No matter how tempted you may be to grab a cut from your favorite CD-audio recording, resist at all costs. Use of copyrighted music, even for internal purposes, is illegal. Copying music is not the same as making a copy of the movie you rented at the video store (that is illegal as well, but enforcement is negligible). In a business context you never know who may be listening and, unlike the average home video viewer, your company may have very deep pockets. There are no gray areas here—just don't do it.

If you are using original music, be certain you own all performance and reproduction rights. With a clip media collection, check to see that the works you are using are royalty-free for your purposes. The packaging should clearly state your usage rights. Keep a copy of the rights agreements and maintain good records noting where you sourced your material. Should you decide to pay for rights to copyrighted materials, the negotiations and contracts should be left up to the legal department.

Audible Overkill

This book defines *cacophony* as: "The unrestrained use of sound and music by an overzealous business presenter." Far worse a sin than failing to take advantage of the persuasive powers of audio is creating a presentation in which music and sound effects hit the audience like incoming mortar rounds. Any media type, when first explored, has a tendency to be overdone. In adopting sound techniques for presentations, it is critical that you err on the side of too little rather than too much. Consider the appropriateness of the effect carefully; sound can be just as powerful at tuning out your audience as tuning them in.

Case Study:

Company: The Lee Apparel
Co., Merriam, Kansas
Business: Apparel
manufacturing
Objective: Develop a
multimedia point-of-sale system
to attract and win over
customers to its line of boys'
apparel.

One of the most timeless and perplexing questions of human existence is, "How do you hold the attention of a pre-teenage boy for more than 15 seconds?" For the giant apparel maker Lee Co., the answer is sound and video.

Young male shoppers, 8 to 12 years old, who visit R. H. Macy's store in Manhattan are drawn into the boys' department by a large, color-splashed display that holds three touchscreen monitors. In the welcome mode, the top monitor displays a full-screen, full-motion video animation of the letter "L," while both the middle and bottom monitors feature an animated "E." The letters are composited against rapidly changing background animations.

The display plays a continuous loop of "hot" and trendy music to grab the attention of potential customers. Once the boys have been lured to stop, prompts on the monitors invite them to touch the screen. When the top monitor is touched, it displays a video presentation consisting of a commercial-like, 30-second montage of boys jumping and running and generally having a great time to the tune of original music

created for the kiosk. The sequence includes slow-motion and other effects, as well as black-and-white text graphics with slogans such as, "Lee jeans are cool."

A second video featurette, which appears when the bottom monitor is touched, has proven to be a big hit with the boys. In it a kid complains that he and his friends are too much alike and that he wants to look different. At that point he magically begins to metamorphose into a variety of different characters, including a dog and a little girl, before "morphing" into himself again, this time outfitted in Lee apparel.

Recognizing that a couple of 30-second TV commercials are hardly enough to hold the intended audience, Lee equipped the middle monitor with an interactive sequence in which the boys can enter their height and weight using an animated tape measure and scale. The program asks if the information is correct, then displays the appropriate jeans sizes. All of these actions are accompanied by an odd assortment of sound effects that provide humor and entertainment.

Lee Apparel Co.

Both the animations and the interactive size/fit program were created by Kansas City–based Spinnaker Communications. The animations, originally hand-drawn by a freelance graphic artist, were scanned into a Macintosh Quadra 950 and redrawn by Spinnaker artists using Adobe Illustrator. CoSA After Effects was used to animate the illustrations, which were then saved as QuickTime movies and printed to Sony Betacam SP videotape. From videotape, the animations were encoded to MPEG digital video format. The two video segments created by Lee were also encoded to MPEG. All of the pieces were then ported to a Philips CD-I 605 CD authoring system running OptImage's MediaMogul software.

According to Kevin Mitchael, Director of Visual Marketing for Lee Co., the project is serving as a test to determine the extent of the company's involvement in CD-I (compact-disc-interactive) and other digital interactive media. "We feel interactive point-of-purchase displays will be everywhere in a few years," says Mitchael. "The possibilities for delivering information with this medium are unbelievable." Mitchael adds that the kiosk is being well received so far by both customers and Macy's personnel.

Sound Basics

In 1979, the sci-fi horror movie *Alien* ran a successful ad campaign with the line, "In space no one can hear you scream." The statement is absolutely correct, physics-wise. Sound cannot travel in the airless vacuum of space (never mind the fact that screaming is not possible either). Sounds that we hear, including screams, are produced by rapid fluctuations in a pressurized atmosphere. When anything—the string of an instrument, a bee's wings, or your vocal chords—moves in the atmosphere, it displaces air molecules. The moving air reaches our ears as sound waves and is translated by the mechanical processes of the inner ear into nerve pulses that we can distinguish and interpret. In essence, the devices used to capture, manipulate, and play back digital sound are analogous to what happens in the human ear.

To make a recording of a sound, an electronic device called a *transducer* converts the sound waves into corresponding changes in electrical voltage levels. A microphone is an example of a transducer. The voltage levels, referred to as an *analog signal,* are then used to create a pattern on magnetic tape (or grooves on a record). For playback, the signal is read off the tape by playback heads, amplified, and passed through a loudspeaker. The loudspeaker is another transducer that performs a conversion opposite that of a microphone. The speakers recreate the original air pressure fluctuations to make the sound audible to our ears (see Figure 9-1).

Sound waves travel at varying speeds, or frequencies. *Frequency* refers to how often the wave completes a cycle in one second and is measured in hertz (Hz). Figure 9-2 depicts a sound waveform of two cycles per second and compares it to eight cycles per second. We detect the frequency of sound as *pitch*. The more cycles per second, the higher the pitch. As an example:

High pitch = High Frequency = Mickey Mouse
Low pitch = Low Frequency = James Earl Jones

The human ear can detect frequencies from about 20Hz to 40Hz at the low end of the spectrum up to about 20,000Hz or 20 kilohertz (20KHz). As people age, they gradually lose the ability to hear frequencies above 15KHz to 16KHz.

The other key measure for sound, and one that should be understood when producing audio for a presentation, is amplitude. Simply, *amplitude* is the perceived loudness of sound that we measure in decibels (dB). The

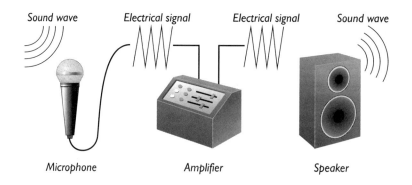

Sound wave   Electrical signal   Electrical signal   Sound wave

Microphone   Amplifier   Speaker

Figure 9-1   Analog signal processing

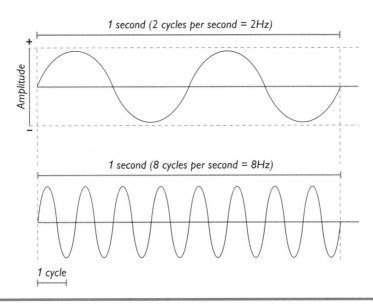

Figure 9-2   Sound waveforms of 2Hz and 8Hz

more pressure exerted by the moving object, the louder the sound and greater the decibels. The background noise created by an audience is roughly 40dB, while a rock concert can generate up to 120dB, which is generally considered the threshold of pain (not to be confused with the rock group Threshold of Pain). Perceived loudness decreases about 6dB every time you double your distance away from the sound source.

Analog sound recording and playback methods date back to the invention of the phonograph by Thomas Edison in 1877 and, while storage methods have changed from rolls to records to tape, there was no fundamental change in the way that sound information was captured and stored until the advent of digital recording.

Digital Audio

Digital recording does the same thing to sound that it does to text or any kind of information. It converts the information to numerical combinations that can be stored, retrieved, and manipulated by a computer.

The digital "plasticity," as you will see, makes the recording and playback of sound much more precise and flexible than analog recording. Adding sound effects narration or music to a presentation does not require that you understand all the theory or technical operations involved; that can be left to the audio engineers. But you should have a basic grasp of the principles of digital audio when it comes time to make decisions about such things as file types, file sizes, and quality levels, as well as hardware and software selection.

The first thing to understand is that in digital recording the changes in electrical values are measured, or *sampled,* by a device called an analog to digital converter (ADC). The ADC records the samples as discrete numeric values. This numerical representation of the audio waveform can then be converted back to an analog signal by a digital to analog converter (DAC) for playback (see Figure 9-3).

To better understand how this works, look at Figure 9-4(a). The wavy lines represent a simple audio waveform. The horizontal line running through the middle of the wave represents the midpoint of the signal strength. Thus, any part of the signal that is at the line is at zero amplitude (volume). Zero amplitude equals silence.

As mentioned, the ADC takes snapshots along the vertical lines at regular intervals of time. This is called the *sampling rate.* The sampling rate is measured in samples per second and typically ranges from 11,000 to 48,000 samples per second (this measure of frequency is also referred to as hertz). In Figure 9-4(b), the sampling rate is denoted by vertical

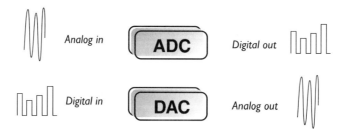

Figure 9-3   Analog to digital conversion and digital to analog
               conversion

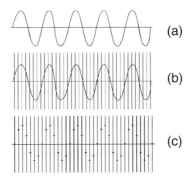

Figure 9-4   An ADC "samples" the analog signal at selected
intervals to create a digital profile of the sound

lines. The samples are taken where the waveform and the vertical lines
intersect. The result is a series of data points, shown in Figure 9-4(c). The
distance of each point from the midpoint is assigned a value. That value
is stored in the file.

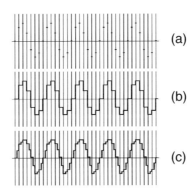

Figure 9-5   A DAC reconstructs the analog waveform with the
aid of filters

To play the audio back, the DAC reconstructs the waveform using the blueprint of the numerical sampling. In Figure 9-5(a) you see a series of data points from a sample recording made for this demonstration. Figure 9-5(b) shows how the DAC connects the dots to reconstruct the original wave. Compare Figure 9-5(b) with Figure 9-4(b) and you will see that the shape of the wave is similar to the original analog wave but not exactly the same.

The angularity of the line has been added to help you visualize how the process of digital to analog conversion works. In actual operation, the sound hardware uses filters that smooth the waveform back to its original shape. The angular wave shows that using a higher sampling rate in the original recording will provide more data points and, therefore, more accuracy in the recording. For example, if the wave were recorded at twice the sampling rate of that shown in Figure 9-4(b), the output would look something like Figure 9-5(c); notice how much smoother the waveform is. Once again, in practice, filters smooth the waveform more than depicted in Figure 9-5.

Digital Advantages

The digital method of recording, combined with compact disc or hard disk storage, offers a number of advantages over the analog tape method of recording and playback. One of the most obvious advantages is the lack of unwanted noise. The magnetic media passing over the head of the tape player/recorder produces an audible *hiss* that can degrade the quality of the sound. Old or low-quality tapes produce even more of the annoying, extraneous noise. Digital recordings add no hiss during storage or playback. Digital filtering circuitry can also be used to "clean up" analog recordings, by eliminating the numerical information that corresponds to the unwanted sound.

As with other forms of digital media, a key benefit of digitized audio is its *random access*: sound stored digitally on magnetic disk or optical disk media allows you to skip forward or back to a desired selection within thousandths or millionths of a second. Locating a selection or start point on audio tape, on the other hand, requires physically advancing or rewinding the reels. Of course, vinyl records are also randomly accessible. (Anyone remember trying to drop the needle down in exactly the right groove so you could hear only the song you wanted?)

When analog sound is transferred from one tape to another, some of the information is lost and some extraneous sound is added. The more dubs or transfers, the more the sound degrades. Digital sound information, if properly handled, can be transferred from one digital medium to another without corrupting the original material.

Accurate positioning or *synchronizing* of a sound is far easier using digital techniques. Unlike analog formats, with a digital recording you can select a location on the disk as precise as a single sample. That represents accuracy in the neighborhood of 20 microseconds (20 millionths of a second). Rarely would any production need that degree of accuracy, but a sound effect that occurs even one-half second early or late is detectable by an audience.

We all get older, but most of us will well outlive the average useful life span of a tape media recording. The tape degrades over time as the magnetic particles lose their orientation, or the tape material breaks down. This causes sound information to be lost, perhaps forever. Because digital recordings are made up of discrete values, they have theoretically infinite longevity. While CD and magnetic disks do ultimately decay, the mathematical information or code does not; transferring the values to new media does not lose or add information. The sound information stays intact unless the code is accidentally erased or corrupted.

Capturing Digital Audio

When recording analog audio, no software is necessary. You simply feed the sound source to the tape recorder, using a microphone or direct connection. Depending on the equipment, you can make adjustments to the recording level by referring to meters and adjusting dials. You can also perform various degrees of filtering. A talented sound engineer can capture audio in the optimal ranges and qualities by knowing what equipment to use for various sources. Note that unless you are using prerecorded sources, you will want to learn as much as possible about good analog recording. The results you get when digitizing will only be as good as the quality of your source recording. The techniques of analog recording are not covered in this book, but dozens of publications can give you whatever background you need.

This chapter assumes you are ready to capture your analog source into digital format. You do that using two primary tools: a sound board (typically the hardware device that contains the ADC/DAC chips) and a sound editing software program. You can digitize virtually any audio source by feeding the signal directly into the sound card through the appropriate ports. For simple voice annotation, you can patch a microphone into the card. Most of the popular editing software programs that let you perform the task of digital capture use the metaphor of an analog tape recorder. At the basic level, they provide buttons for play, rewind, stop, and record (see Figure 9-6). Most programs also have graphical representations of the sound waveform. The better programs also have graphical representations of the recording meters found on analog equipment.

Note that Figure 9-6 has a number of recording options depicted on the interface that must be set by the user before digitizing the sound. Ironically, because of the precision with which it captures and translates sound waves, digital recording can be less forgiving than analog recording. A basic understanding of the following technical procedures and options will help you achieve the best possible results when recording and editing digital audio.

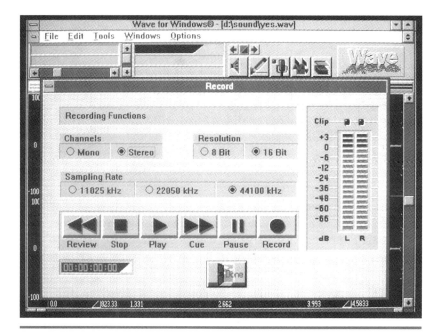

Figure 9-6   Wave for Windows sound editing interface from Turtle Beach Systems

AUDIO SOURCE	SAMPLING RATE (HZ)
Speech	4,000–8,000
Sounds (low pitch)	11,025
Sounds (high pitch)	22,050
Music	22,050–44,100

Table 9-1   Frequency Recording Guidelines

Sampling Rate

For the best recording, you must choose the right sampling rate. According to established audio principles, the lowest sampling rate that can be used to record a particular tone is twice the frequency of the tone. Using a sampling rate lower than this will generate low frequency tones that did not exist in the original material. This phenomenon is called *aliasing*. To ensure that you use the correct sampling rate, you simply take the maximum frequency of the sounds in the audio to be recorded and multiply by two. This is your minimum sampling rate.

Of course, you may not have handy an electronic device to analyze frequencies, in which case you can use your ears and Table 9-1 as backup. The table can be used as a general guideline for the sampling rates to use for certain types of audio. In general, if the audio has primarily low tones, such as an explosion, bass guitar, foghorn, or human speech, low sampling rates in the 4KHz to 11KHz range are sufficient. Audio with primarily high frequencies, such as sneezing, clapping, and music with cymbal crashes and violins, requires much higher sampling rates in the 22KHz to 44KHz range.

The higher frequency sampling rates always sound better but are not always practical, because they consume much more space on your hard drive or other storage medium. The trick is to match the quality of the recording with the available storage space to come up with the best compromise.

tip

Using your sound recording equipment, try recording your voice and some music at 11KHz, 22KHz, and 44KHz. Then listen to the files through good speakers or headphones and notice that the higher sampling rates sound much better, but make less difference for the voice recording. This is because the frequency spectrum for a conversational voice is much smaller than for music.

Sample Size

Depending on the capabilities of your sound equipment, you will be able to record and play back digital audio with a sample size of either 8-bit or 16-bit samples, or both. The more bits per sample, the greater the accuracy of the recording: *8-bit sampling* defines 256 steps in the amplitude of the signal, whereas *16-bit sampling* provides 65,536 steps. Though 16-bit produces better sound reproduction, the greater amount of data creates larger files. So, the decision about which sample size to use becomes a question of tradeoffs between quality and storage.

For most sound effects and voice annotation, 8-bit is fine. For music, though, 16-bit samples are preferable. CD-quality audio is recorded and played back at 16-bits. For voice recognition applications, 16-bit will offer greater accuracy. Once again, let your ears be the final judge. Clip music or sound will also be available either in 8-bit, 16-bit, or both. It is a good idea to have both versions available to meet whatever quality or storage requirements you might run into.

Stereo Versus Mono

Selecting stereo or mono is perhaps the easiest choice you will make. While stereo may seem to be the obvious choice, it is only useful for simulating the position of a sound relative to the listener's position—sounds such as a car passing by or an orchestra playing. If it is not necessary to have a sound seem to come from a particular location in space, you are better off using mono, because mono uses half the hard disk space of stereo.

Recording Level

Good analog tape recorders have a record level adjustment that allows you to set the volume of the signal being recorded. Setting it too low results in a noisy recording, full of hiss. Setting it a little high usually does not make a noticeable difference. Setting it too high results in distortion. Digital recording is different. A limited range of values can be used to store audio. As a result, you are not given the luxury of setting the recording volume a bit too high. If you exceed the maximum allowed by your equipment, you have no digital space left to place the extra information represented by the higher volume. Only the maximum value allowed is recorded and the rest is ignored, resulting in a condition known as *clipping*. Figure 9-7 shows a correctly recorded waveform and the same waveform recorded improperly so that the sound is clipped.

The best digital recording is one that reaches the maximum level but never exceeds it. This is not easy to do, but having a record meter in the recording software makes it possible to get close without going over. The meter is usually a bar that shows the amplitude of the signal as it is picked

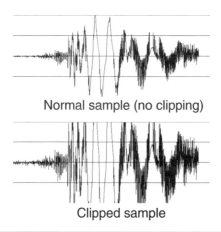

Normal sample (no clipping)

Clipped sample

Figure 9-7   Sample of normal (top) and clipped (bottom) waveforms

up by the computer. With a meter, you can watch the levels of the input signal as the music or sound effect is played. Set your recording level so that at the loudest point in the music or sound effect the meter reaches 90 to 95 percent of the maximum. This will ensure that you do not clip the signal.

Some editing programs allow you to *normalize* the digital data, which amplifies the recording so that the loudest point in the recording is set to the maximum allowable value. If you have access to this function, it will close the gap between your 90 to 95 percent peak and the 100 percent maximum. Figure 9-8 shows the virtual recording level meters in an audio editing software program. Notice that they look much like analog meters found on professional analog recording equipment.

Storing Digital Audio

By now you should be getting the message that the biggest hurdle to overcome with digital sound is the management of memory space. There is no nice way to say it—digital audio is a storage hog. Consider that one minute of music from your favorite CD-audio stored in your computer at the same quality would demand nearly 10.6MB of memory.

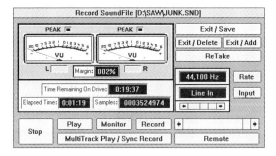

Figure 9-8   The recording meter interface from Innovative Quality Software's Software Audio Workshop (SAW)

Take a look at Table 9-2, which compares ten seconds of audio recording at different settings. Begin by noticing the wide range of storage requirements. A 10-second sound file sampled in mono at 8 bits and 11KHz occupies 110K, while the same sound recorded at 16 bits in stereo at 44KHz will gobble up almost 2MB of storage.

To determine the amount of storage required for a particular recording, use the following formula:

File size(bytes) = *Sampling rate*(Hz) × *Sample size*(bytes) × *Channels*(1 for mono, 2 for stereo) × *Length*(seconds)

Using the formula, a 30-second, 16-bit, 44.1KHz, mono file would occupy 44,100(Hz) × 2(bytes) × 1(mono) × 30(seconds), or 2,646,000 bytes.

SAMPLE SIZE	SAMPLING RATE (IN KHZ)	CHANNELS*	FILE SIZE (IN KILOBYTES)**
8 bits	11.025	1	110
8 bits	11.025	2	220
8 bits	22.05	1	220
8 bits	22.05	2	440
8 bits	44.1	1	440
8 bits	44.1	2	880
16 bits	11.025	1	220
16 bits	11.025	2	440
16 bits	22.05	1	440
16 bits	22.05	2	880
16 bits	44.1	1	880
16 bits	44.1	2	1,760

* Mono = 1, Stereo = 2

**Ten seconds of audio

Table 9-2   File Size Variables

reminder

The abbreviation KHz represents kilohertz, or thousands of samples per second; MHz represents megahertz, or millions of samples per second. And 8 bits is equal to one byte; 16 bits is equal to two bytes.

File Formats

Because multimedia production and delivery are spreading to many different computer platforms, you are likely to encounter different file types in the course of production. To help you identify the origin of a sound file, the following list contains the most common file formats and a short description of their contents. Pay particular attention to the Audio Interchange File Format (AIFF) and the Wave (.WAV) subset file format—these are the files you will run across most often if you are a Macintosh or Windows user, respectively. The list is ordered by the file format name with the usual file extension in parenthesis. Note that there are two types of computer-resident sound referred to here. *Digital audio* is a general term for any voice, sound effect, or music converted from analog form to digital. *MIDI* is the standardized designation for music files generated by digitally controlled musical instruments and devices.

▶ *Audio Interchange File Format (AIFF):* AIFF is used on the Apple Macintosh and Commodore Amiga for storing digital audio. A variety of sampling rates and sample sizes, up to 32 bits, are supported. An additional feature of AIFF is the support of *looping,* the repetitive playback of a block of audio. The file format is also used on UNIX-based machines, such as the Silicon Graphics Indy.

▶ *Musical Insturment Digital Interface, MDI (MID, MFF):* The MIDI file format is an international standard for storing MIDI data (see "The MIDI Dimension" section later in this chapter). As such, it is the perferred and most universal way to share MIDI data. There are two file types: Type 0 and Type 1. Type 1 is a newer format and supports multiple tracks.

▶ *Resource Interchange File Format (RIFF):* The Microsoft RIFF file format can contain many different types of data, including

digital audio (WAV) and MIDI (see the description of the Wave format following).

▶ *Roll (ROL):* The ROL file format was defined by AdLib, Inc. for use with the AdLib line of sound cards. It stores MIDI-like data and Yamaha FM synthesizer information.

▶ *Sound (SND):* This generic file can have ambiguous origins. It is traditionally a digital audio file for the Apple Macintosh that is limited to 8-bit data. However, some programs use the same file extention to signify digital audio.

▶ *Sun Audio (AU):* Sun Audio is a 16-bit compressed digital audio file used by Sun Microsystems' workstations.

▶ *Turtle SMP (SMP):* Turtle SMP files store digital audio data. The file format was defined by Turtle Beach Systems for use with their digital audio recording and editing software.

▶ *Voice (VOC):* The VOC digital audio file format from Creative Technology became popular with the introduction of the Sound Blaster sound card. The file can contain 8- or 16-bit digital audio with or without compression.

▶ *Wave (WAV):* The Wave file format is a subset of the RIFF specification from Microsoft. It is for digital audio storage and supports stereo or mono, 8- or 16-bit audio. Wave supports a variety of sampling rates, but the most common are 11.025KHz, 22.05KHz, and 44.1KHz.

Audio Compression

As is clearly shown in Table 9-2, waveform sound files can quickly grow in size as you raise the quality and increase the length of music and effects. Left unchecked, sound files can easily overwhelm the storage capacity of your system. Unless your use of sound is minimal or your storage capacity gargantuan, you will need a way to reduce the amount of sound data before storage. For example, say you want to run a three-minute song in the background of a montage of text, still images, and animation. You have already used 35MB of the 40MB you have allocated to the sequence on your hard disk. You digitize the song at moderate quality levels of 16 bits at 22.05KHz in mono. The resulting 8MB file puts you well over your storage budget.

Fortunately, technology and standards have been developed to compress audio files before storage and decompress them before being played. You may already be familiar with compression technology for data files. Documents and graphics are frequently compressed before being transferred over networks, online, or on disks. Compression works by using algorithms to mathematically interpret the data and reduce the amount of code necessary to represent the information. The algorithm is applied in reverse to decompress the file. The resultant savings in space is typically 2:1 or greater. In the preceding example, compression would reduce your three-minute music file to 4MB or less.

Compression Schemes

Two basic types of data compression can be used: lossless and lossy. As the name implies, when using *lossless* compression, no data is lost and the compressed file will sound exactly the same as the original. Where precise maintenance of data is required (compressing mathematical formulas, for example), lossless must be used. *Lossy* compression, on the other hand, causes the file to retain less information than it had to begin with. The resulting playback will have degraded proportionally to the amount of data that is lost.

It may seem that under no circumstances would you want to use a lossy compression scheme when storing your audio files. Yet, lossy compression algorithms yield significantly higher compression ratios than lossless ones. In practice, good compression algorithms using lossy formulas will cause minimal audible quality loss. Also, lossy compression and decompression is usually easier for the computer to process. Using a lossy method gives the computer more free time to perform other tasks during playback. This can be of particular importance if you are processing animation or video at the same time.

tip

When using lossy compression methods, do not compress, decompress, then recompress audio files. Each step will further deteriorate the quality of the audio, because lossy compression loses some audio data each time.

The most common methods of lossy audio compression are Adaptive Differential Pulse Code Modulation (ADPCM), A-law, and μ-law. ADPCM is by far the most popular compression standard and is used by numerous hardware- and software-based compression products. ADPCM yields compression ratios from 2:1 to 10:1 (ratios beyond 4:1 are not generally adequate for music compression) by measuring the change in an amplitude of the values between successive samples, then using the results to predict the next values. In effect, the formulas take an educated guess at where the amplitude is going. The measurements are taken so frequently there is very little error in the predictions, and the sound emerges virtually intact.

A-law and μ-law compress audio at a ratio of 2:1. Up-and-coming algorithms, such as Audio MPEG and specific algorithms for speech, can compress digital audio efficiently at more than 20:1.

Compression can be accomplished with either hardware or software. Hardware compression takes the burden of processing the algorithms off the computer's central processor, but it is more expensive. Some sound boards feature hardware-based compression. Others bundle in software-based compression programs. Compression programs can also be purchased separately.

An alternative way to keep music files short is to use MIDI instead of digital audio. Though the file sizes vary based on content, MIDI files are roughly one-hundredth the size of comparable waveform digital audio music files. MIDI is discussed in detail later in this chapter.

Digital Audio Editing

Using digital sound for multimedia presentation requires that your personal computer is equipped with the necessary ADC and DAC chips, as well as amplifiers, filters, and other circuitry. Typically, these elements are combined on a sound board or sound card inside the computer. (Sound boards are discussed in more detail at the end of this chapter.) To record and edit your sound, you will also need sound capture and editing software.

At least two dozen editing packages exist on the market today. Microsoft Windows-equipped computers come with a basic audio editor; most sound boards also bundle audio editing software. The software varies in quality and capability in rough proportion to price, which goes from about $50 to $1,500. The simplest ones do little more than cut out unwanted bits, tighten up passages, adjust amplitude (referred to as *gain* in some programs), and insert fades. Intermediate-level packages add features such as special effects (echo, reverse, reverb) and filters for cleaning up extraneous noise. The most advanced products give you exceptional control over the recording process, as well as the use of soundtracks for professional-style editing.

The level of software you require will be determined by how often you work with sound, which types of sound elements, and how much technical control you need. As a rule, the more features, the more you pay. Unless you are a professional musician or sound engineer, you probably would not need to spend more than $300 (probably less) on an editing package. The software accompanying your computer or sound board might also be adequate for your needs.

In fact, the best way to begin learning audio editing is to use the software that comes with your system. Become familiar with the range of functions and features and the basics of audio editing techniques; then, if you require other capabilities, seek out a package that fits your production needs as well as your budget. Be certain you purchase software that supports the sampling rates and sample sizes of your sound board and vice versa. Also, make sure it will support the range of sound file formats you will be using with your multimedia authoring software.

The following simple techniques will introduce you to the principles of audio editing. Once you understand them, you will be able to use at least the basic features in just about any editing package you select. You can practice the techniques using the software that comes with your audio card. Or you can use WAVE for Windows, available on the CD-ROM that accompanies this book.

Editing Techniques

As explained, digital audio files are made up of discrete values representing the amplitude of an audio wave at specific points in time. This data can be represented graphically as a series of tight vertical lines arranged along a horizontal axis. Audio editors use this graphical method to visualize the recorded signal for purposes of cutting, pasting, transposing, adding special effects, and other operations. Figure 9-9 depicts the left and right channels of an audio waveform (or amplitude) timeline for the sound of breaking glass. The interface, here designed for

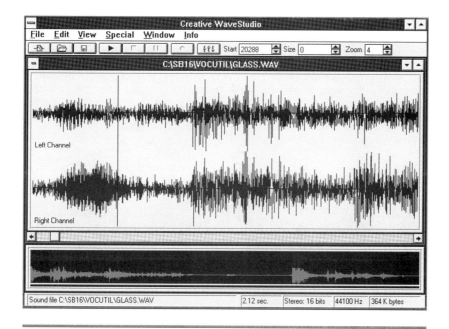

Figure 9-9   Creative Wave Studio waveform editing interface
from Creative Labs, Inc.

use in the Windows environment, is typical of interfaces found in most
editing software.

Here is the waveform (or amplitude) timeline of the spoken line,
"What do you want to do?"

The longer the lines, the higher the amplitude. Note that it is relatively
easy to tell where one word leaves off and another begins, because of the
natural silence that falls between most words when a person is speaking.
You can also instantly see which words are loudest and where the peak

gain is in each word. All of this information comes into play any time you perform an edit.

The most common sound edits are simply cutting, pasting, and copying. These actions are similar to the cutting and pasting actions when using text or graphics, and the tools operate much the same way. To cut out a piece of sound, you first define the area you wish to remove by highlighting it with your mouse or keyboard commands. Here, the waveform for the words "do you want" is highlighted:

Pressing the cut, delete, erase, or equivalent command in the audio editor eliminates the segment from the waveform:

The sentence would now say, "What to do?" You can reinsert the segment in the same place or a different location by clicking where you want it to go and using the paste, insert, or equivalent function in the editing software. Here, the clip is returned to the file, but at the end:

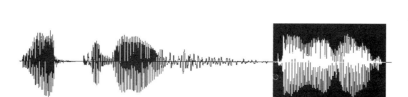

The sentence would now say, "What to do do you want."

You might desire to repeat a sound segment, either for emphasis or as an attention-getting special effect. To create a repeat, perform a copy function. Some editors require that you select copy and then paste the copy where you want it. Other programs let you perform the copy without using the paste step. Here, the final segment has been copied once to the end of the file:

The copied portion is still highlighted. If you were to play the file now, it would say, "What to do do you want do you want."

In essence, audio editing is that simple. Learning to add special effects and perform filtering operations can take some time to master, but most presentations will not require much tinkering. Clip music, for example, may need to be shortened or lengthened with the appropriate fades added. Or, you may want to clip out a particular segment. Sound effects often need to be cleaned up at the beginning and end

Airworks

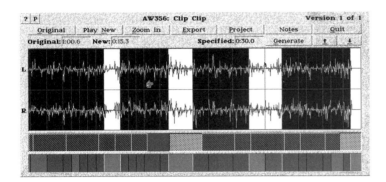

One of the most perplexing problems producers face when adding music to a presentation is finding a clip that precisely matches the duration of the visual it is designed to accompany.

Cutting the music off abruptly at the end creates a harsh and unprofessional transition. A new program, TuneBuilder from Airworks, combines a specialized audio editing pro-

gram with professionally coded clip music to let presenters edit music to specific lengths with just a couple of commands. The music can be cut and pasted at the beginning, at the end, and at specified segments throughout the piece. The program then reassembles it to create a new clip. The program outputs to all sound file types. A few royalty-free music collections that have been customized for TuneBuilder are available now from third-party vendors. More are on the way.

to get rid of extraneous noise or to make them the right duration for the visual they accompany.

With narration, you may want to remove long pauses between words, or get rid of unwanted ums and ahs. On occasion, you may even want to alter the sense or meaning of a phrase to change its effect on the audience. For example, here again is the phrase, "What do you want to do?"

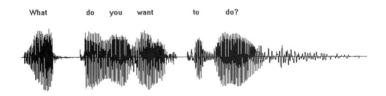

This time, the word "what" is cut from the beginning of the file and pasted at the end of the file. Next, "do" that now opens the phrase is completely deleted. Finally, the little bit of dead space between the words "want" and "to" are eliminated to make the sentence sound more natural:

What used to be "What do you want to do?" is now "You want to do *what?*"

The MIDI Dimension

In addition to recording and editing waveform digital audio, a multimedia PC or Macintosh gives you access to the musical world of MIDI (pronounced "middy"; see the "File Formats" section earlier in this chapter). MIDI was developed in 1983 as a standard to allow digital electronic musical instruments and computers to communicate with each other regardless of the brand of the device by putting music into a digital form that represents not only the notes to be played, but also what instruments should play them, how loud, and for how long. The device that creates the original MIDI information is called a *controller*. Electronic keyboards, drum pads, and guitars are all examples of MIDI controllers. Even your computer keyboard can function as a MIDI controller if the right software is installed.

To hear the sound represented in the MIDI file and generated by the controller, a synthesizer must be added to the system. The *synthesizer* "reads" the MIDI data and generates the corresponding music. In many cases, as with most electronic keyboards, the controller and synthesizer are combined in one device.

The information generated by the controller can go directly to a synthesizer for output with amplification, or it can be stored as a MIDI file in the computer. A computer equipped with a MIDI-capable sound board can then play back the file to an onboard or external synthesizer.

The critical thing to remember about MIDI is that, unlike a digital recording, which contains actual sound consisting of thousands of samples for each second of music, MIDI simply defines the instruments and notes that are to be played and how they should be played. The synthesizer is given the responsibility of creating the actual sounds. MIDI instructs synthesizers how to play music, something like a conductor and sheet music instructing the musicians in an orchestra how to play a symphony.

Because they contain only the notes and related instructions, and not the actual music, MIDI files are many times smaller than digital audio files of roughly the same duration. As previously noted, this option can be of tremendous advantage when strategizing storage space for a multimedia presentation.

Many sound boards have MIDI capability. As such, they can be connected with an external controller or even use the computer keyboard to create MIDI files. The synthesizer on the card can then play back the MIDI information.

Connecting MIDI devices

External MIDI devices generally have two ports, In and Out, for receiving and sending MIDI data, respectively. When connecting these devices, MIDI cables are attached from the Out of one device to the In of the other device and vice versa (see Figure 9-10). At first, it might seem logical that devices should be connected with matching ports, In with In and Out with Out, but just the opposite is true. The best way to ensure that the cables are connected correctly is to remember that MIDI messages coming *out* from the port of one device must go *into* the port of another device. The setup is analogous to recording from one tape deck to another.

Some devices also have a port marked Thru. This port simply duplicates all MIDI messages that arrive on the device's In port. As a result, anything attached to the Thru port will receive the same MIDI message as the device to which it is connected. The Thru port is used to chain several MIDI synthesizers to a single controller's Out port.

Figure 9-11 demonstrates how the MIDI data coming out of device #1 arrives at the In port of device #2. That data is copied to the Thru port and sent to device #3. Now both devices, #2 and #3, will see the same MIDI messages and will be able to play identically.

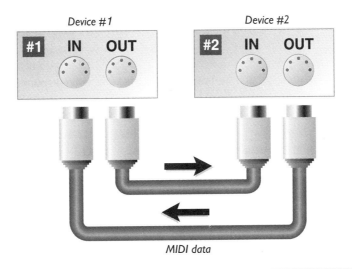

Figure 9-10   Correct patching for MIDI devices

Recording and Storing

After the controller generates the notes and other information, and before the data can be sent to a synthesizer for output, the data is processed by a sequencer. *Sequencers* function much like word processors

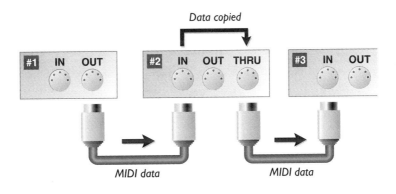

Figure 9-11   Correct patching for MIDI devices

for music: the MIDI information is arranged and prepared for recording, either by hardware or software. Some keyboards have built-in sequencing and recording, but sequencing is usually done with the aid of a computer. Sequencing software lets the computer store the MIDI data into a MIDI file for later playback and editing. It also allows you to display the notes onscreen, either as traditional musical notation, or as bars in a table that look similar to the holes in a roll of player piano music (see Figure 9-12). In fact, the way a player piano roll triggers the piano keys is a good model for thinking about how MIDI works.

As mentioned earlier, because a MIDI system does not record the actual sounds in the digital file, its storage requirements are significantly less than digital audio. Unfortunately, no straightforward formula exists for computing the size of MIDI files. Usually, about 6K per minute is a good estimate. Five minutes of MIDI music, then, requires about 30K of disk storage, compared to more than 50MB for the same duration of CD-quality digital audio (16-bit, 44KHz, stereo). The file size is a function of the complexity of the music and the number of instruments

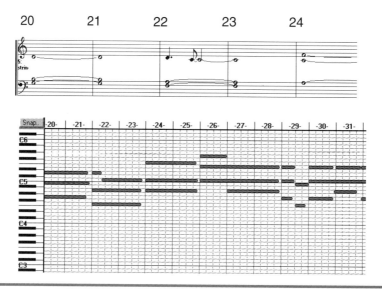

Figure 9-12   MIDI data displayed as traditional musical notation (top) and in tabular notation (bottom)

indicated. Heavy use of effects, such as pitch bend and aftertouch, which send a lot of messages, will add dramatically to the size of the file.

MIDI can also improve the performance of the multimedia computer system by requiring less data to pass from the hard disk to the sound card. CD-quality audio, for example, requires that 176.4K of data travel from the hard disk to the sound card for every second that the sound is played. Throughput can put serious limitations on sound playback, sometimes causing gaps or delays to the amplifier and speakers. Reducing the volume of data, as MIDI does, reduces the throughput demands on the internal architecture of the multimedia system.

Digital Audio or MIDI?

Digital audio and MIDI both offer a different set of advantages and disadvantages. Fortunately, no law says you cannot use both on your system or even in the same presentation. The points listed below will help you differentiate between them, based on what you want to accomplish, and help you adopt the format that best fits your needs.

▶ *Consistent audio quality:* Because MIDI relies on a synthesizer to play back the encoded music, playback quality will vary according to the quality and characteristics of the synthesized sounds. The timbre and tone of the MIDI-generated instruments can vary drastically. Digital audio, by contrast, holds all the music information and plays it back exactly as it is in the file, ensuring that the quality of the music in your presentation will be consistent on every machine.

▶ *Compatibility:* Since digital audio is recorded at a specific sampling rate and sample size, you must be sure that the computer you use for your presentation is capable of playing back your sounds at the recorded sampling rate and sample size. For this reason, you must sometimes record at the expected lowest common denominators for the potential playback system(s). MIDI is not affected by either sample size or sampling rate. It will play back exactly as intended on any MIDI-capable machine and, as mentioned, is dependent only on the synthesizer for playback characteristics.

warning

MIDI is not widely used on the Macintosh platform. For that reason, MIDI is not generally viable for multiplatform delivery of presentations.

▶ *Sound effects:* MIDI does not record or play real-world sound effects, but it can be used to synthesize them to some extent. Only digital audio allows you to play any kind of sound—a door slamming, the boss screaming, the other shoe dropping, or whatever else you want. If you can record it, you will be able to play it in your presentation.

▶ *Flexibility:* Since the entire musical score is resident in a MIDI file, you edit virtually everything about it—instruments, individual notes, timing of notes, tempo, and so on. And you can do your edit on the file without ever returning to the controller. You do not have this kind of flexibility with digitally recorded music unless you re-record.

▶ *Small file size:* Again, MIDI files are considerably smaller than digital audio files of the same duration. If you have limited hard disk space or you want to be able to fit your presentation on a floppy, MIDI is an ideal way to do it.

▶ *Low processor impact:* A lot less processor power is needed to play MIDI files. If you are using a slow computer for presentations, you may not have the horsepower for digital audio. Even if you have a fast machine, using MIDI for background music will leave more processor headroom for digital sound effects and animations.

tip

Bottom line on MIDI and digital audio: If you have a lot of hard disk space or master to CD-ROM, digital waveform music sounds great. Otherwise, limit digital audio to sound effects and MIDI for your music.

Audio Hardware

Standard PCs usually do not come with a sound board, but PCs meeting the MPC standards do (see Chapter 3 for more on MPC standards). Most Macintosh computers have either 8-bit or 16-bit sound processing built in. A few computers now come with sound processing built directly into the motherboard of the CPU—this is a trend that can be expected to continue as sound processing becomes more important to all types of computer applications. For laptop presenters who do not have built-in audio or the necessary board space, external sound boards are available, but they are mostly limited to 8-bit performance.

Sound boards vary widely in cost, features, and bundled software. In general, the more costly boards have high-performance features, such as MIDI, digital signal processing (DSP), audio compression, voice recognition, wavetable synthesis, and more. Most boards on the market can play back at 16-bit quality. Many low-end boards, costing as little at $70, only record at 8 bits. Sound boards with similar capabilities can sound remarkably different, and you should try to listen to a few before you buy; but generally, they can be judged for quality on their component specifications.

The key factors are sample size (8-, 10-, 12-, or 16-bit) and sampling rate (usually 11-, 22-, 44.1- or 48KHz). The higher values in both produce the best sound. As previously stated, however, the higher rates use more memory space, so a good board will have software that allows you to select the desired levels. For multimedia playback, a sound board with 8-bit recording and 16-bit playback will be sufficient. For multimedia authoring, a sound card should have both 16-bit recording and playback capability. You will also want a card with an FM synthesizer, which produces effects for video games and uses the MIDI control language to mimic musical instruments.

tip

The difference between a sampling rate of 44.1KHz and 48KHz is undetectable to all but the most sensitive ears. You do not need to pay a higher price for the higher sampling rate.

Look for a sound board that handles a variety of sound sources, such as a tape deck, CD-ROM, microphone, and other outside stereo sources. You should be able to mix multiple sources with control over sound levels and channel inputs. The PC platform offers the largest selection of sound cards by far. Creative Labs' Sound Blaster 16, Media Vision's ProAudio Spectrum 16, and Cardinal Technologies' Digital Sound are three examples in the $200 to $300 price range that provide 16-bit digital audio capability and at least moderate quality FM synthesis. To get wavetable synthesis for the highest quality MIDI music, you can opt for boards that are in the $300 to $600 range, such as the Creative Labs Sound Blaster AWE32, Turtle Beach Multisound, and Advanced Gravis Ultrasound MAX.

Sound add-ons for the Macintosh have been a nonissue from the beginning, because Macs come with 8-bit audio capability built in. Some of the newer Macs, such as the Quadra 660AV, have built-in, 16-bit digital audio. As a result, most wave cards for the Mac are intended for high-end use. If you are in the market for more audio power than what comes with your Mac, boards such as the Digidesign Audio Media II give you the power that only recording studios could afford just a few years ago. For most multimedia purposes, though, the sound that is built into the Mac is more than enough.

Speaking of Speakers

Speakers are a technological afterthought for most multimedia-capable computers. The small (very small) acoustic satellites that sit at the sides of the monitor are typically the least expensive items in the system. Inferior speakers are one of the biggest mistakes you could ever make in a multimedia presentation. After all that work converting to digital, editing, synchronizing, and effecting, it should be a capital crime to play back through any device that will diminish or mask the quality of the source material.

Because the evaluation process tends to be so subjective, speakers are a touchy subject. Just what constitutes good reproduction? How much speaker is enough? Everyone can agree that some speakers sound better than others. Given that limitation, here are just a handful of characteristics you should keep in mind when looking for a pair of speakers for multimedia development and presentation.

One of the first things to decide is whether or not to buy self-powered speakers. If you are building a multimedia studio, you might have the luxury of feeding your audio signal to a full-blown stereo system. In most cases, though, your system will not be connected to a pre-amp, amplifier, graphic equalizer, or other components. Instead, your speakers will rely either on their own power or the meager wattage that comes out of the sound board. It is fairly easy to discount sound board–powered speakers, unless that is all your budget will allow. They do not have either the frequency range or the power to do justice to a multimedia soundtrack. Powered speakers, by contrast, have improved dramatically in recent years, thanks to the push toward multimedia. Powered speakers have a built-in amplifier; you plug them into the sound card, attach the power cord to the wall, and you are ready. Some speakers are battery powered, but the chance of a speaker going dead at the wrong moment makes them a gamble to use.

warning

Speakers placed close to your monitor should be magnetically shielded. If they are not, the warping and color distortion will permanently damage your monitor. Unshielded speakers can also erase your hard disk. If you are not sure what kind of speakers you have, keep them far away from your computer.

A variety of companies make powered speakers, but the most prominent are Labtec, Roland, Acoustic Research, Altec Lansing, Sony, and Bose. Prices range from $30 to $500 or more. Usually, the more you pay, the better the sound, but this in not always the case. The more expensive systems usually give you extra features, such as tone control, multiple inputs for mixing, and more power.

No matter what price range you are looking into, sound reproduction capability is what you are after. The ideal human ear is capable of hearing frequencies in the range of 20Hz to 20KHz. As a result, speakers capable of producing sounds over this range are desirable. In reality, most people's hearing is limited in the upper range to 18KHz at best, even lower for people over the age of 60. Look for speakers that at least reach 18KHz.

On the low end of the audio spectrum, sounds in the 30Hz and lower range are generally felt more than heard. An explosion, for example, contains a multitude of low frequencies that you cannot really hear but that make the windows rattle. For maximum low range reproduction, buy a system with a subwoofer (see Figure 9-13). The subwoofer is designed with a large speaker that only produces low frequency sounds. This has the added advantage of allowing the satellite speakers to remain relatively small, as they will be producing only high frequencies. You do not need two subwoofers to reproduce stereo effects, because low frequencies are not directional. Another advantage: you can place the subwoofer anywhere in the room; the human ear cannot easily detect the direction of origin of low frequencies.

tip

Many subwoofers have crossover frequencies above 100Hz (i.e., the subwoofer must reproduce sounds above 100Hz), which can diminish the effect of stereo sound because only low frequencies are nondirectional. Your ear can locate a single subwoofer generating higher frequency sounds.

One final feature to look for when inspecting the frequency response of speakers is the balance of frequencies. Look for speakers that have a flat response over their claimed frequency range. All frequencies within the range should be equally loud. Some speakers do a better job at this than others.

Finally, you should decide how much power you will need. Most sound cards can provide four to eight watts of power per channel (left

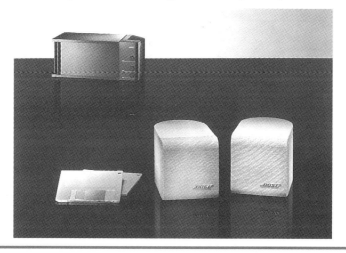

Figure 9-13   Bose Acoustimass-3 Multimedia Speaker system uses a separate subwoofer and two satellite speakers

Sound Test

Perform this test to find speakers with the desired flat response. First, select a few speakers that appear to have good specs and sound good to your ears. Then, play a noise source through the first pair; the noise should be something like what you hear when you tune between stations on the radio. Connect one pair of speakers and listen to the noise, paying attention to any frequencies that tend to stand out. Try the next pair of speakers and you will notice that they sound different. The frequencies that they favor will almost certainly be different. Even the best speakers tend to sound louder at some frequencies. Look for those that favor particular frequencies the least.

and right). For an office, this might be sufficient; but for a small group in a boardroom, you will need 10 to 20 watts per channel. Most powered speakers provide at least this much power; some provide 30 watts or more per channel. If you plan on giving a presentation to a larger audience, a good loudspeaker or P.A. system will be necessary. They should provide more than 100 watts. A high-powered system may go up to 1,000 watts of power, but at that point your audience may be able to stay home and still hear the presentation.

Sound Decisionss

Everyone has different opinions about the importance of sound. The variations stem from differences in culture, profession, and physiology. You may be among those who have never thought much about sound, its properties, and its power to influence. The lesson to be learned here is that even if sound does not seem to be having much of an effect on the conscious, intellectual level, its effect is being felt below the surface.

The easiest way to learn to use sound effectively in multimedia presentation is to become a discerning listener. Learn to focus on sound and to detect changes in yourself and the audience. Soon you will begin to think aurally. As a result, you will have opened a new dimension for persuasion and communication.

CHAPTER 10

Things seen are mightier than things heard.

—Alfred Lord Tennyson

VISUALLY SPEAKING: WORKING WITH VIDEO

Almost since its invention, video has been a powerful corporate communications tool. For many organizations an in-house video department is an expensive necessity, for others, only an item on a long wish list. Digital video technology is changing that picture, making video not only more affordable for companies of any size, but easier for the nonpro to learn, use, and integrate into multimedia presentations.

The Story in Motion

Motion enthralls us: a jet streaking across the sky, a cat leaping from a roof, an acrobat tumbling. Perhaps the instinct harks back to our years in the primordial forests when we learned to detect movement as an alert to predators or prey. Or, perhaps we are irresistibly drawn to movement by the electronic glissando triggered in our brains by our tracking eyes. Whatever the organic underpinnings, dozens of studies confirm that motion grips us and communicates to us in ways that stimulate interest, heighten comprehension, and improve retention.

In business applications video has the power to

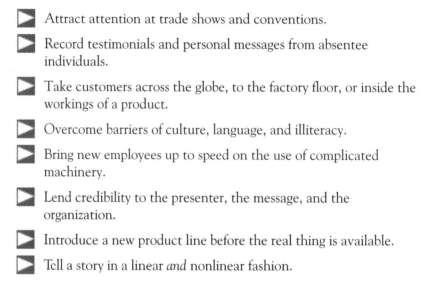

▷ Attract attention at trade shows and conventions.

▷ Record testimonials and personal messages from absentee individuals.

▷ Take customers across the globe, to the factory floor, or inside the workings of a product.

▷ Overcome barriers of culture, language, and illiteracy.

▷ Bring new employees up to speed on the use of complicated machinery.

▷ Lend credibility to the presenter, the message, and the organization.

▷ Introduce a new product line before the real thing is available.

▷ Tell a story in a linear *and* nonlinear fashion.

In short, video has the capacity to communicate the most amount of information in the least amount of time with the greatest impact. When used correctly, video provides audiences with information they would not have been able to access or understand as well in any other way.

Evaluating Video

In business, video is too often used inappropriately or used simply because it is available. Before committing to digital video in your presentation strategy, first determine if video is the best medium for the message. Evaluate your options: Would a technical illustration tell the story best? Could what you have to say be said better using still images with audio? Would the diagrammatic nature of animation convey the information more effectively?

Of the five basic media components—text, graphics, animation, audio, and video—video places the greatest demand on time, resources, and development and delivery systems. Accordingly, you should discipline yourself to use video only when it is clearly the best way to make your point. In general, video excels in the following four basic areas of presentation.

Describing Motion

Motion, by its very nature, is difficult to put into words and is rarely communicated in still photos. If motion is integral to the content of the message, video is an obvious choice. For example, it might take pages of description to detail the cornering capabilities of a new car model when 30 seconds of video from the test track would say it all.

Demonstrating Procedures

Video allows the accurate demonstration of tasks in context. The trainee or customer sees the tools or equipment being used in the actual environment. Video, as part of a multimedia presentation, can be controlled interactively, letting the presenter or the user slow down, pause, stop, reverse, skip ahead, or restart until the procedure is learned and fully understood.

Conveying Emotion

Video excels at demonstrating human emotional and psychological reactions and interactions. It can convey the subtleties of facial expressions or the heights of enthusiasm that would otherwise be lost with still images and sound bites. Well-produced, emotive video segments can accelerate product recognition, increase viewer awareness, and facilitate acceptance of ideas and concepts.

Simulating Situations

Video can re-create real-world or hypothetical events. It can help to teach cause and effect relationships by depicting scenarios with real

Case Study:

Company: First Data Corp., Health Systems Group, Charlotte, N.C.

Business: Data processing and information services for 2,000+ client base

Objectives: Develop a high-tech presentation for use when selling information solutions to health care organizations.

For First Data Corp., a high-tech image is more than just desirable, it's crucial. The worldwide information systems company specializes in high-volume data processing—from credit card statements to MCI operator assistance to healthcare information. Needless to say, its clients must be convinced of First Data's mastery of information technology. So, when the Health Systems Group needed a presentation to help sell its health-care information systems, multimedia was an obvious choice.

"We're developing high-tech healthcare information solutions, so we should be doing high-tech presentations—not slide presentations or, heaven forbid, overheads," explains Paulette Wilder, First Data's manager of marketing communications.

First Data commissioned the Whitley Group, a Charlotte, N.C.–based multimedia production company, to create a sales presentation that could be used in the field by its sales force, at trade shows and by First Data's corporate executives.

The presentation, authored in Macromedia Director, begins with a welcoming screen pairing a spinning First Data Corp. logo with the prospective client's company name—the client name is entered by the salesperson prior to the presentation and appears throughout the presentation.

The next sequence transports the audience into the telemedical world of the future as the salesperson holds up what looks like a universal health card, passes it over the screen, and clicks a button. A picture of the ID card, complete with the salesperson's name and photo, comes up on the screen. Text flashes next to the image and an electronic voice says, "Please repeat name for voice recognition."

This sequence was made possible by incorporating the names and photos of the 50 or so sales and marketing personnel who will be using the presentation into a database within the program. Prior to a pres-

entation, the individual presenter clicks on a drop-down menu and selects his or her name from the list. This action triggers a command that brings up the presenter's likeness on the simulated universal health card.

After the mock-identification process, the salesperson/patient selects a "consultation" option and enters into a simulated dialog with a physician. The physician appears in a small video window as if from a live videoconference link. The doctor displays the patient's chart and X-rays while discussing the case, then phones another doctor for consultation. As the virtual visit comes to a close, the consulting physician makes a tentative diagnosis and schedules an appointment for the salesperson/patient.

The extensive use of Quick-Time video, combined with the other graphical elements, is intended to give the viewer the feel of an actual videoconference in progress. Additional video testimonials from satisfied customers and industry experts help to bolster the company's credibility. Videos of executives familiarize the prospective client with the company style and key contacts in an effort to establish an early rapport.

First Data

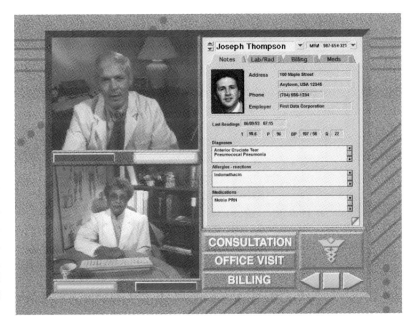

The nine minutes of video used in the presentation were shot with a Sony Betacam SP video camera by a freelance videographer. The video was then captured digitally with a Radius VideoVision board and Apple's QuickTime utility running on a Macintosh Quadra 800. Initially, the video segments alone occupied 100MB of storage space. The files were reduced to roughly 60MB using SuperMac Cinepak compression software. The video was then edited using Adobe Premiere. Even with compression, the final presentation takes up a whopping 100MB of hard disk space. The video runs in a 240×180 window, at 10 frames per second.

The QuickTime implementation presented a few challenges, admits Eddie Thompson, multimedia project manager for the Whitley Group. All of First Data's salespeople were using Apple PowerBook Duo 270C portable computers, a machine with good power and speed, but like most off-the-shelf laptops, not optimized for video and audio playback.

According to Thompson, the capture process sometimes caused the video and audio elements to lose sync. As a result, he had to resync the audio and video during the editing process using Adobe Premiere. In addition, when the final presentation was loaded from an external drive onto the hard drive of First Data's PowerBook Duo docking station and then onto each of the Duos, all of the links for each QuickTime movie had to be reestablished, Thompson says. Because of the different hard drives, he explains, it is necessary to respe-cify the location of each QuickTime movie on the new hard drive.

The video in the final product jerks a bit and the dialog is slightly out of sync, but the overall effect communicates First Data's cutting edge capabilities and presents a persuasive message: Our company can keep your company on the cutting edge as well.

people in realistic situations. Managers, for example, can be taught to handle difficult employee problems by watching actors play out a scene. Using interactive techniques, managers can not only be asked to suggest a course of action, they can immediately see the results of their suggestions.

Analog Video

Analog video is a mature and widespread technology. All broadcast and cable transmissions initiate from videotape sources; the ubiquitous VCR is in more than 80 percent of television-owning homes; and sales and rentals of movies on home video are nearly three times the revenues from movie theaters. Video as a presentation tool dates back to the 1960s, and it is being used today more than it ever has to achieve any and all of the presentation goals outlined earlier. While the technology has improved dramatically during nearly four decades, the basic analog process and components have remained relatively unchanged (see Figure 10-1).

Analog video requires a camera, a recording deck, and a playback deck—playback and recording decks are often one in the same. Many camcorders combine all three functions in one self-contained unit. Very

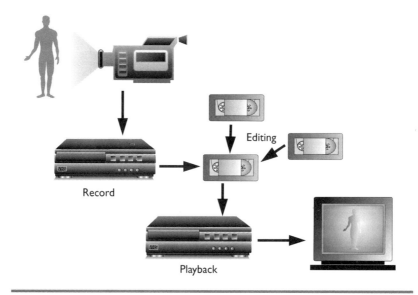

Record

Editing

Playback

Figure 10-1   The analog video process

simply, the process begins when the video camera focuses light through a lens at timed intervals (shutter speed). The light intensity is detected by one or more *CCDs* (charge-coupled devices). CCDs are 1/2-inch microchips arrayed with 380,000 photo-sensitive diodes. The diodes generate an electric charge proportional to the light intensity that strikes them. The charges are scanned, converted to electronic signals that represent the scene, and sent to the recorder. Magnetic heads in the recorder create magnetic patterns on the videotape as it passes. The video signal is read from the videotape by a magnetic playback head, which then sends the signal to a monitor for viewing.

Videotape is edited by recording a section from one tape onto another. The clips are arranged in order, with the desired effects, transitions, and titles added in. A sound track is typically recorded with the video images and can be synchronized using any of several systems, the most widespread being the *SMPTE* (Society of Motion Picture and Television Engineers) time code.

As with analog audio, the video signal is composed of modulated frequencies or electronic pulses that are analogous to the original moving images. The video signal can be of various qualities and configurations, depending on what equipment and standards are used in the recording process. In the U.S., Canada, and Japan video is recorded, edited, and played back using the *NTSC* (National Television Standards Committee) format. The *PAL* (Phase Alteration by Line) format is used in the U.K., Germany, China, and Australia. In France, the format used is *SECAM* (Système Electronique pour Coleur Avec Mémorie).

The main point to understand about NTSC is that it transmits video images at a standard 30 frames per second (fps)—a speed deemed optimal for creating the illusion of motion. In comparison, film-based motion pictures are projected at 24 fps. Analog video is stored in such a way as to be played back at a resolution between 250–750 horizontal scanlines. NTSC, it should be noted, is capable of displaying considerably more information than is received over the air waves as television. The 6MHz (including sound) bandwidth limitation of television broadcast frequencies allows only about 360 lines to be carried to the receiver.

The horizontal lines that make up one frame, or one picture, are displayed in two fields, each containing only every other line of information. The 60 *fields* per second yield a frame rate of 30 *frames* per second. These 60 fields and 30 frames per second are played back on an NTSC format monitor using a technique called *interlacing*. Interlacing rapidly displays the two adjacent fields to give the impression of a single frame. The key factors to remember, as you begin to learn about digital video, are that analog NTSC video displays an interlaced signal at 30 fps and about 525 lines of resolution. The rest you can leave to the video engineers.

The PC Meets Video

The marriage of video and the personal computer for presentation purposes dates back more than a decade. The first computers used to control tape decks were proprietary devices specially designed to streamline and automate the editing process. Later, tape decks were equipped with an RS232 port so they could be hooked up to standard personal computers. Software running on the PC could be used to control the record, playback, and advance/rewind functions of the deck, both for editing and for playback during a presentation or training session. Because of the complex programming, most implementations of computer-controlled video playback were reserved for high-end presentations or permanent installations. Museum displays are one example.

The most serious drawback of videotape as a presentation medium is its linear nature. Videotape-based presentations must run from start to finish, or to a designated stop point, then resume at the same point. Jumping to discontinuous spots in the tape requires that the tape deck be advanced or rewound, causing an unavoidable time lapse in even the fastest machines. For video editing purposes the wait is tolerable, for presentations and training just the opposite.

The linear limitations of videotape were first conquered with the invention of the videodisc or laserdisc. (LaserDisc is a Pioneer Electronics tradename that has come into general usage as laserdisc.) Videodiscs, which can hold up to 30 minutes of high-quality video and CD-quality audio on one side, are 12-inch diameter plastic platters that are read by a laser. They physically store information in much the same way as a CD-ROM does, the difference being that videodiscs store the video as FM signals and CD-ROMs store the information as bit data.

Videodisc players equipped with the necessary circuitry and interface can be controlled with single-frame accuracy for use in a variety of presentation and training applications. In contrast to videotape decks, videodisc players offer fast random access to video and still images (54,000 stills per side). Unlike videotape, which can be erased and altered, once information is written to a videodisc it cannot be changed—discs that cannot be altered are said to be "read-only" or "write once" discs.

The video information on tape and videodiscs is stored in analog form and will only display on an analog monitor or television screen. The same signal will not display on a computer monitor without additional hardware—the use of a video overlay card lets video be shown on a selected area of the computer screen. Videodisc systems, which continue to be widely used in training, kiosks, and educational applications, range from simple linear playback of analog video with no interaction to sophisticated systems that allow full control of the deck by an external computer. For presentations that require random-access NTSC video, they continue to be a useful presentation tool. But despite videodisc's successful history and spectrum of applications, the format is slowly, inexorably being muscled aside by digital video's inherent strengths in cost, portability, convenience, and flexibility.

The Digital Video Revolution

The introduction of videotape in 1956 revolutionized the use of "moving pictures" in entertainment, news, and business communications by allowing moving images to be stored, edited and played back in a far faster and less costly manner than film.

As happened with sound recording, discussed in the previous chapter, digital video evolved from a combination of analog recording methods and computer technology. Digital video information is stored as binary code in exactly the same way a computer stores still photographs, illustrations, or any other type of information. The difference between video and static images, however, is that the video image is time based. The video frames—each one a separate image—must be played back at a constant, rapid speed to achieve the illusion of motion. In addition, sound accompanying a digital clip must be digitized and stored in such a way that the two data streams stay synchronized as the information is fed from the digital storage device to the video and audio processors and from there to the monitor and speakers.

Once converted to digital form, the existing video is far easier to edit and enhance than in its original videotape format. Clips of video can be "cut and pasted" to form new segments. Transitions and effects can be added to enhance the shift from one scene or subject to another. Audio can be inserted, deleted, or updated. It is also possible, and relatively easy, to add titles, graphics, and animations.

Overall, the dramatic improvements in cost reduction and convenience introduced by digital video technology are not unlike the advances that analog videotape production offered over film-based motion picture production nearly 40 years ago.

Digital Advantages

Digital video, when properly implemented, offers a number of benefits in presentation development and delivery. Some, such as cost savings, are hard-line advantages. Others, such as the potential for increased interactivity, are "soft" benefits, but just as important to understand.

Cost Savings

Digital video solutions on the personal computer have dropped rapidly in price and at the same time have become easier to configure and use. Assembling a digital video system that will give you NTSC-quality video comparable to videotape or videodisc is still a costly proposition, but even those technical obstacles are falling quickly. Broadcast-quality digital video boards are already available for less than $10,000. Look for that figure to drop rapidly as the competition heats up. Keep in mind, too, that the most significant cost savings will be felt in the reduced editing time, easier changes, and far lower duplication costs of digital video.

At the lower end of the scale, where broadcast-quality production is not a necessity, the entry level price has fallen below $1,000. With an inexpensive capture board, readily available software, and a camcorder (or any video source), you are ready to capture, compress, and play back video clips that can be incorporated into presentations, spreadsheets, and even text documents.

Enhanced Interactivity

Videotape's linear nature makes it inherently slow to access. As such, videotape-based delivery systems can provide only very limited interactivity. Videodisc systems offer random access to information, but tend to have relatively slow access times on all but the most expensive players.

Digital video presentations, particularly those that run on powerful PCs with fast hard drives, can deliver almost instant random access, thereby avoiding distracting pauses and long wait times.

Cleaner Duplication and Editing

Analog video quality degrades rapidly with each generation because of unwanted "noise" from the tape and the recording heads. Even if you begin with the highest possible tape quality, after several duplications in the editing or distribution process, the image will be far less sharp and clean than the original. To lessen this effect, all editing is typically done on copies, then the selected cuts are recorded to a master from the source tapes. If changes are made later, a new master must be recorded to maintain the quality level.

Because digital video information is stored as numeric values, copies of the data contain exactly the same information as was first captured. Theoretically, a digital video clip can be reworked indefinitely without quality loss. In practice, it should be noted, digital video image quality does degrade when it is compressed and decompressed. Compression technology causes information to be lost or extraneous information to be added. It is important to limit the number of times video is compressed, decompressed, and recompressed. As is explained in detail later in this chapter, if the project requires relatively high video quality the use of hardware-based capture and compression equipment can greatly reduce the loss of information.

Storage Flexibility

Analog video is time based and can only be stored at the full screen size in which it is captured. Consequently, each clip of analog video requires the same amount of storage regardless of the manner in which it is used—if an analog clip is played back as a small window on the screen, it still must contain all the signal information for a full screen. Digital video, by contrast, can be captured and stored in the particular size for which it is intended in the final presentation. For example, if the video is to fill only a quarter of the screen, the image can be captured and stored as 320×240 pixels. Digital video, in this way, saves not only storage space but the inherent cost of storage as well.

Malleability

Digital video outshines analog video in its ability to let the presenter make changes quickly and easily. With digital video, a core presentation can be created as a template or model that can be used for any number of later presentations as well. The information in the original is simply dropped out or changed and new information added. The overall structure and design is retained, saving both the cost and time of creating a new presentation from scratch.

Platforms

Before you get too much further into the detail of digital video, be sure that you understand the platform requirements. This book assumes you will be creating interactive multimedia presentations that will incorporate digital video, not linear-only videotaped presentations. It also assumes that you will be developing and delivering your presentations on Apple Macintosh or IBM-compatible machines, or both. You will need a computer system that is properly configured and powerful enough to handle the demands of digital video. Later in the chapter you will learn about the specific PC requirements for different levels of video performance and different types of video peripherals. For now, keep in mind that you may end up dealing with two entirely separate computer systems— one for developing and another for delivering your presentation. In almost all cases, the best results will be achieved when the two systems are identical, especially on low-end systems that do not provide hardware assistance for compression.

The Mac

Because so few industry-wide standards exist for multimedia, you can never be certain how well your presentation will run on other machines unless you have direct control of both the development and the delivery platforms. The problem is especially critical for digital video. The Macintosh comes close to a standardized platform by incorporating its own and the industry's standards for such factors as video display resolutions, digital audio, networking, media integration, SCSI support, CD-ROM, and digital video. Macintosh computer models such as the Quadra 660AV, Quadra 840AV, and PowerPC with the AV option offer integrated NTSC video support. Since only minor changes have been made over the years to Macintosh standards, it is possible to deliver digital video presentations developed today on Macintosh computers that are four or more years old.

One reason the Mac has dominated multimedia development is that during the early days of multimedia production technical limitations allowed for information to travel from the Macintosh to Windows machines but not the other direction. When creating presentations to be delivered in both the Macintosh and Windows environments, developers had little choice but to work on the Mac. Recently, these barriers have begun to break down. Software developers have addressed the problem by building bidirectional, multiplatform compatibility into their authoring packages (see Chapter 5). Today, you can play your Mac-based presentation on PCs, and with the right software you can use your Macintosh to play back presentations that were created on IBM-compatible machines.

The PC

In contrast to the Mac, PCs have been a standards wasteland, particularly in the area of digital media integration. In recent years, vendors have made attempts to provide some standards guidance to the PC platform by forging the MPC (Multimedia Personal Computer) specifications (see Chapter 3). In an attempt to include as many products into the specifications as possible, the standards process has focused on the lowest common denominators for multimedia. The first MPC standards were quickly eclipsed by MPC-2; an all-new MPC-3 specification is in development.

As the standards evolve, the digital video requirements for delivery machines change. For multimedia production, the shifting of standards makes it difficult, if not impossible, to know what level of machines will be in use at any given site. For example, many highly rated, high-performance video (monitor) display cards for the PC will not work with digital video.

tip

Unless you have no other option, don't rely on the stated capabilities of a delivery machine, or even the MPC-level specification. The only way to be certain the digital video will run properly in your PC-based presentation is to put everything together and test it.

QuickTime and Video for Windows

Unarguably, the biggest breakthrough for personal-computer–based digital video was the development of Apple's QuickTime movie format. QuickTime is a standardized file format for production and playback that allows video and audio to be captured and combined in a file on one machine and played back together on any other machine running Quick-Time. While QuickTime files can only be created on a Macintosh computer, they can be played back on a PC running Apple's QuickTime for Windows. Following Apple's lead, Microsoft introduced the Video for Windows AVI (Audio-Video Interleaved) format. Video for Windows also combines audio and video for standardized production and playback. AVI files created in Windows can be converted to the QuickTime file format.

Both QuickTime and AVI support multiple types of compression—the process necessary to reduce file sizes to a manageable level. Both formats also *interleave* the audio and video. Interleaving writes the audio information and the video information together on the file so that when played back they appear to be in sync, even when video frames are dropped.

The Contribution of Compression

While static images such as photographs and illustrations translate relatively well to the PC from their native formats, video does not. With static images, the computer processes and displays one screen of organized pixels at a time. Each screen is complete in itself—a separate information event. The presenter controls the interval between images as desired. Video, on the other hand, is a time-based medium that requires rapidly changing images (as many as 30 per second, but not less than about 10) to be strung together at consistent speeds. This presents a tremendous challenge to the processing power of a personal computer.

Key to understanding the dynamics of using digital video is an awareness of the sheer size of stored video information. Uncompressed digital video files are the King Kongs of the computer world, the behemoths of bytes. One second of full-motion video can consume as much as 30MB. A single minute would use 1.8GB of storage space, or the equivalent of 18 100MB hard drives.

Obviously, to make digital video practical, the file sizes must be reduced using some form of compression technology. The hardware and/or software devices that do this are called *codecs* (short for *compression/decompression*). The process for compressing video files is similar to that of compressing audio discussed in the previous chapter, though many times more complex. There are dozens of algorithms—the mathematical rules that govern specific software tasks—that can be used to accomplish the necessary compression. Most can be divided up into two types: *symmetric* and *asymmetric*.

Symmetric Compression

Symmetric compression can be simply understood as any algorithm that allows video to be compressed (captured) and decompressed (played back) in the same amount of time. This compression process is often referred to as *real-time encoding and decoding* because it occurs at the speed at which the video plays. The advantages to real-time compression in both directions are obvious. During development the compression occurs as the video clips are being played into the system. There is no time lost waiting for the processor to finish its job. The principal drawback to symmetric compression is the demands it puts on the system. Because of the massive data rate, symmetrically compressed video plays back far better on a hardware-based system.

Software-based symmetric compression programs include *Apple Video* (standard with QuickTime), Intel *Indeo*, and *Microsoft Video 1* (standard with Video for Windows). All have slightly different characteristics, but can accomplish real-time capture and playback without the need for additional compression hardware. Indeo, a compression file format based on Intel's DVI (Digital Video Interactive) technology, allows video compressed with hardware to be played back with software only. Intel uses this approach with its Smart Video Recorder capture board.

Hardware-based symmetric compression is accomplished with the aid of microchips that have been programmed with specific compression algorithms. Using hardware to compress and decompress digital video files greatly improves the performance of the entire system.

JPEG (Joint Photographic Experts Group) is a hardware-based standard that was developed for use with still images. It is referred to as an *intraframe* codec because it works by reducing the amount of digital information in each frame. This is also called *spatial* compression. Maintaining a large amount of information with JPEG results in relatively large files; giving up information for the sake of file size causes relatively high distortion. Also, the standard does not apply to file structure, so there is no guarantee you can play back a file compressed with one JPEG board on a board from another manufacturer.

Asymmetric Compression

Asymmetric compression was developed to address the need for good quality digital video at a low data rate. Lower data rates allow digital video to be delivered easily from CD-ROM or over a network. Asymmetric algorithms work by letting the computer have as much time as it needs to capture, compress, and store the original video. Complex mathematical calculations must be performed on each frame of the video, so the process of compressing and storing each frame cannot be completed in real time. At the extreme, some asymmetric compression schemes take as much as 300 times longer to process a frame as it takes to play the frame.

SuperMac *Cinepak* is an example of a software-based, asymmetric codec. It produces compact files, but requires an hour or more to compress one minute of video.

The *MPEG* (Motion Pictures Experts Group) codec is an asymmetric, hardware-based approach to reducing video file sizes that works by detecting the differences between frames and then eliminating the storage of redundant information. Because it looks at key frames and predicts changes between them it is referred to as *interframe, frame differencing*, or *temporal* compression. File compression cannot be accomplished in real time, but the file sizes are significantly smaller allowing for higher quality at larger image sizes.

DVI, another hardware-based codec, is a proprietary system from Intel Corp. that combines some of the best features of the JPEG and MPEG standards. It produces exceptional video results, but requires that a special DVI chipset such as IBM ActionMedia II be incorporated into the video board.

Table 10-1 can be used as a general guide to determine which combination of symmetric and asymmetric, as well as hardware and software-based codecs, will be most appropriate for your application.

CAPTURE	PLAYBACK	PROS AND CONS
Software-based symmetric *Apple Video* *Intel Indeo* *Video for Windows*	Software-based symmetric *Apple Video* *Intel Indeo* *Video for Windows*	**Pros:** Fast development time, inexpensive. **Cons:** Poor quality capture and playback. Quality limits playback to smaller window sizes. Demanding on delivery CPU.
Software-based symmetric *Apple Video* *Intel Indeo* *Video for Windows*	Software-based asymmetric *Cinepak*	**Pros:** Better quality video playback than with symmetrical compression alone. **Cons:** Longer development cycle, minimal increase in playback quality. Poor capture quality. Window sizes limited by processor speed.
Hardware-based symmetric *DVI* *JPEG*	Software-based symmetric *Apple Video* *Intel Indeo* *Video for Windows*	**Pros:** Better quality video capture than with software capture. **Cons:** Fair quality playback. High data rate increases demands on delivery CPU. Capture hardware expensive.
Hardware-based Symmetric *DVI* *JPEG*	Software-based asymmetric *Cinepak*	**Pros:** Good quality. Can deliver to multiple computers on CD-ROM. **Cons:** Longer development cycle. Expensive capture hardware. Window size limited to 320×240.
Hardware-based symmetric *DVI* *JPEG*	Hardware-based symmetric *DVI, JPEG*	**Pros:** Fast development. High-quality capture and playback. Up to full screen. Faster editing. DVI over network or CD-ROM. **Cons:** Expensive add-in card for delivery machines. Limited installed based for delivery.
Hardware-based symmetric broadcast-quality analog video compressed by service bureau	Hardware-based asymmetric *DVI, MPEG*	**Pros:** Video rivals or surpasses videodisc. Playback boards much cheaper than DVI. **Cons:** Access to expensive analog video editing equipment. Service bureau compression costs. Expensive DVI conversion.

Table 10-1   Codec Capture and Playback Configurations

Capture and Playback

When determining the settings for digital video capture and desired results for playback, there are four main factors that need to be balanced in order to achieve the most effective results:

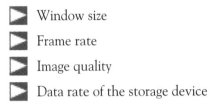 Window size

Frame rate

Image quality

Data rate of the storage device

As you will see in the digital capture and editing tutorial that follows, all the variables play an important role in establishing settings for compression.

You should begin the capture process with the last item on the list, determining the rate of data transfer, the *data rate*, of your storage device. This is the speed at which the storage device sends the file information to the processor. The faster and smoother the data transfer rate, the smoother the motion effect. Hard disks are the fastest devices available, delivering a data rate of about 1MB/second up to about 10MB/second (transfer rate is discussed in more detail later in this chapter). Obviously, even the fastest drive falls well short of the 30MB/second needed to process one uncompressed second of digital video. Transfer rates for other playback devices are even slower. A dual-speed CD-ROM, for example, has a data transfer rate of only about 300 kilobytes per second (k/sec)— the effective rate is often lower, about 200 k/sec.

The *window size* represents the number of pixels that will be displayed horizontally and vertically, respectively. The smaller the window, the faster the information can be processed. The typical window size settings are shown in Figure 10-2. Note that 640×480 represents full screen; 320×240 is not half screen, but quarter screen. 240×180 fills one eighth; 160×120 is only one sixteenth of the screen area. Obviously, the smaller windows made necessary by the limitations of digital video force you to pay a price in the design of the presentation and what kind of video images you include.

A full-size 640×480 window can be achieved either with actual pixels corresponding to those dimensions, or by *interpolating* a 320×240 window.

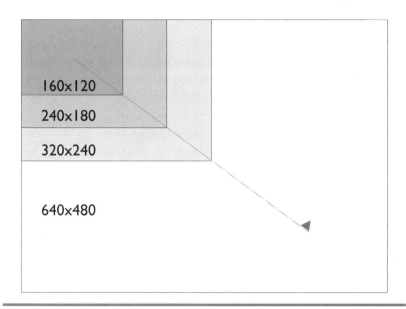

160×120

240×180

320×240

640×480

Figure 10-2 Digital video window sizes

Interpolation, in effect, duplicates each pixel, thus filling up twice the screen area. The resulting image will, of course, be of lower quality than an actual 640×480 screen.

note

A broadcast quality, NTSC image contains slightly more pixels (704×485) than the VGA (640×480) computer standard. During capture, a small part of the NTSC image at the edges may be lost.

Frame rate varies from an extremely slow 7 to10 fps to full-motion 30 fps. Figure 10-3 depicts one sixth of a second and demonstrates how a lower frame rate reduces the amount of information that must be processed and displayed. The tradeoff at lower frame rates comes in the form of stilted or jerky motion. Below about 10 fps the illusion of motion is gone and the video appears more like a series of still shots. Not until you reach between 24 and 30 fps does the noticeable "silent film" flicker disappear.

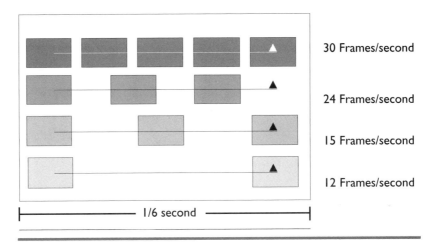

30 Frames/second

24 Frames/second

15 Frames/second

12 Frames/second

├─────────────── 1/6 second ───────────────┤

Figure 10-3   Frame rates

Image quality refers to the amount or bit depth of digital information that the codec (compression/decompression) device will capture in each frame. The settings can range from low-quality 256 (8-bit) capture in which only 25% of the full range of picture information is stored, to 100% or 1,024 (24-bit). The highest quality corresponds to a full-color, high-resolution still image. Reducing the level of image quality reduces the amount of information that must be processed and vice versa. See Chapter 8 for a description of bit depth.

The four diagrams in Figure 10-4 depict in graphic form a fixed CD-ROM data rate (or bandwidth) of 300 k/sec at different settings of window size, frame rate, and image quality. You can see that since a limited amount of data can be processed at any given time it is necessary to strike a balance between the other settings to reach the optimum video quality. In Figure 10-4(a), a large window size (320×240) will tend to consume bandwidth very rapidly, reducing the frame rate to 15 fps and the image quality to 50 percent. What you get is somewhat herky-jerky, quarter-screen video of only moderate image quality.

In Figure 10-4(b), a maximum frame rate is chosen (30 fps) but the window size must be drastically reduced to only 160x120 with the image quality that will be moderate. Figure 10-4(c) depicts a typical setting for digital video in the QuickTime and AVI formats. The window size is still

Figure 10-4 Digital video capture and playback variables

small, the frame rates are slightly better, and the image quality is quite good.

Finally, Figure 10-4(d) depicts a well-balanced trade-off between the three factors. Indeed, it is the solution most commonly supported in products that are delivered from CD-ROM. Note the reasonable window size (240×180), a modest frame rate (15–24 fps), with moderate-to-good overall quality (50 to 75 percent).

System Requirements

Asking a computer's central processor to add digital video to its already long list of processing functions is a bit like throwing a busy juggler a half dozen more balls. Even if the CPU is able to keep up, performance will suffer. The speed of the delivery machine's CPU, in fact, will be the greatest determinant of digital video performance. Using, at minimum, a 486 processor running 33MHz in a PC or a Motorola 68040 processor running at 25MHz on a Mac Quadra will give you small-but-workable window sizes: 240×180 at 15 fps (frames per second), 320×240 at 30 fps using the newest version of QuickTime.

Adding secondary or multiple secondary processor hardware to a system to take the burden off the CPU can dramatically increase the quality, size, and speed of your video. Because of this shared responsibility, digital video can be run on a system with less processing power than would be required of a software-based system. Hardware-based digital video requires a CPU of at least a 486SL at 25MHz for the PC or a Motorola 68030 processor running at 16MHz (Macintosh IIci). Playback-only cards with Intel's DVI video processing technology can be purchased for as little as $1,200 and include the ability to play back CD-quality digital audio as well.

Because hardware-based digital video systems allow much higher capture quality than software solutions, hardware-based systems are almost always going to be the desirable option for video capture. At the very end of the development cycle it is possible to convert the high-quality, hardware-based video to a variety of software-only formats for delivery. Inexpensive hardware playback solutions are under development. MPEG, for example, provides full-screen, full-motion, high-quality video on almost any computer at an additional cost of only $300–$500.

Producing Video for Multimedia

The elements involved in the production of digital video fall into roughly four stages. Each stage plays a significant role in the quality of the overall results.

1. *Analog video acquisition.* Includes selecting the analog equipment and shooting the video—the analog footage that will later be digitized.

2. *Digital capture, processing, and storing.* Includes selecting hardware and software for compression and decompression.

3. *Editing, adding effects, and titling.* Includes selecting the editing software and special effect add-ins.

4. *Delivery and display.* Includes the sound boards, speakers, CD-ROM, and video scan converters as well as display devices.

Analog Video Acquisition

Before you incur the time and expense of a video shoot, you should search around a bit to see what might already be available. As you will see in this chapter, virtually any existing video material can be converted into a digital format and "repurposed" for use in your presentation. In some cases, as with commercially available clip video, your material may already be digitized. Here are some places to look:

▶ *Existing videotape or videodisc presentations.* You may find exactly the footage you need in a presentation that was delivered by someone else in your organization. The subject may have been entirely different, but the video clip might be just what you need.

▶ *Television commercials.* The tapes from your company's advertising could contain professionally filmed clips showing a new product or service.

▶ *Technical and training videos.* Engineering or design departments within your company may have created clips or entire productions that can be edited to fit your presentation. Check with the training department to find out what they have in their training archives.

▶ *University and public libraries.* Many private and public resource centers have large bodies of work that include public domain video material.

▶ *Stock footage houses.* Video stock footage companies can supply a wealth of material, for a fee. The material will not be specific to your company, but can be useful for demonstrating principles, adding historical or social context, setting the mood, or injecting humor.

▶ *Clip media vendors.* Generic video clips relating to business and just about any other subject are available on CD-ROM from more than a dozen companies. Working samples are available for your use on the CD-ROM that accompanies this book.

Video Hardware

Analog video hardware is available in four tiers of performance and price: *consumer, prosumer, industrial,* and *broadcast* (see Table 10-2). Within each category, you will find a variety of features and, in some cases, a wide range in prices. Keep in mind when selecting a camera or tape deck that

	CONSUMER	PROSUMER	INDUSTRIAL	BROADCAST
CAMERAS				
Tape formats	VHS, 8mm	S-VHS, Hi8*	S-VHS, Hi8*	3/4-inch, 3/4-inch SP, Betacam, Betacam SP, MII, D2
Imaging chips	Single CCD	Single CCD	Triple CCD	Triple CCD
Connectors	RCA	RCA, BNC, S-Video	Composite, BNC, S-Video	Composite, BNC, S-Video
Resolution	230–270 lines	300–350 lines	400–450 lines	700–750 lines
Features		LANC Control	SMPTE	SMPTE
Price	$400–$900	$1,200–$1,800	$3,500–$7,500	$10,000–$65,000
VIDEO DECKS				
Tape formats	VHS, 8mm	S-VHS, Hi8	S-VHS, Hi8	3/4-inch, 3/4-inch SP, Betacam, Betacam SP, MII, D2
Signal to noise ratio	42dB	48dB	55dB	58dB
Connectors	RCA	RCA, BNC, S-Video	Component RGB	S-Video, BNC, Component RGB
Features	Hi-Fi	Digital effects	SMPTE	SMPTE
Price	$200–$500	$900–$1,800	$5,000–$8,000	$8,000–$65,000

Table 10-2 Analog Video Hardware

the quality of the digital video capture will ultimately depend on the quality of your original image. That does not mean, however, that buying the highest quality equipment on the market is always justified. The performance level of all the components in your multimedia development configuration will determine what level of performance is required for capture. For example, if the final delivery is 160×120 QuickTime or Video for Windows, it matters very little what camera you use to record the video. In this case, the techniques you use to set up the shot and overcome the small window size will matter far more.

Most PC-based presentations use a 240×180 window; Macintoshes allow up to a 320×240 window. As a rule of thumb, you would want to use at least prosumer-grade products and high-quality S-VHS or Hi8 videotapes for shooting. Consumer devices should be used only when the development system cannot offer a level of quality that takes advantage of the extra resolution offered by prosumer and industrial equipment.

Once information is recorded to videotape with a *camera*, a *video deck* may be required to play the information in order to digitize the video. In some cases, the camera used to capture the video can be an effective playback unit as well. The range of quality, performance, and features of video decks are similar to those for cameras and fall into the same general categories.

tip

If the situation allows, try recording straight from the camera into the capture card of the computer without taking the information to tape. This will often double or triple the apparent quality of the image. This is a good way to effectively use consumer and prosumer equipment for high-quality capture.

Consumer

Basic consumer-grade *cameras* are widely available through retail and specialty stores. They typically support VHS or 8mm video formats and offer a video resolution of about 250 horizontal lines, or less than half of the maximum possible resolution for NTSC video. The devices are almost exclusively designed around a single low-resolution CCD (charged coupled device) imaging system and offer the bare minimum of quality needed for serious development. The consumer-grade devices also offer very few custom exposure controls to adjust for difficult lighting conditions. Prices start below $500.

Consumer-grade *video decks* are widely available, inexpensive, and offer comparable quality to consumer-grade video cameras. The market is dominated by VHS recorder/players, but a few 8mm models are available. In general, the quality of these devices is not good enough to reproduce a useful video image for computer-based capture because of the inherent low resolution. Prices for consumer decks start at less than $200 and go to about $500.

tip

Do not invest in a top-end consumer model. It makes more sense to spend a bit more to buy a lower-end prosumer deck.

Prosumer

In recent years a family of prosumer-grade *cameras* has been developed to fill the gap between consumer and industrial equipment. Prosumer devices can cost two to three times that of similar consumer gear, but they offer higher resolution and more features. The cameras and recorders in this class support multiple tape formats, either VHS and S-VHS or 8mm and Hi8. The S-VHS and Hi8 formats offer higher-quality recording than conventional VHS or 8mm tape. In addition to having added quality, the prosumer devices typically sport a range of control options that are found on professional equipment, including controls for shutter speed, fades, white balance, gain, and more. Most of the prosumer cameras and recorders still only have a single CCD imaging system, but the ability to use the higher-quality tape lets you capture with improved resolution—around 350 horizontal lines. Prosumer devices range in price from $1,200 to $1,800.

Prosumer *video decks* are often sufficient for good-quality video capture (see Figure 10-5). Recently, these decks have been embellished with features found only in industrial or professional equipment. They commonly support time code and methods for computer control. Prosumer decks also accept the high-quality videotape formats S-VHS and Hi8. Prosumer decks, in general, are built to withstand harsher handling and tend to have a longer useful life. Prices start at just less than $900 and go to more than $1,800. The extra cost throughout the price range is usually associated with tape formats, number, and quality of video inputs and outputs, or more features that may or may not have an effect on the overall quality of the final video capture.

Figure 10-5   Panasonic AG-7750H S-VHS Hi-Fi editing deck

tip

Never use a video deck that is lower in quality than the camera that was used to capture the video. For example, it would not be advisable to use a consumer-grade video deck to play back video shot from a prosumer or industrial camera. In this the case, use the higher-quality camcorder itself for playback.

Industrial

Industrial-grade video *cameras* offer significantly better quality than their consumer or prosumer counterparts (see Figure 10-6). While industrial cameras mostly use the same S-VHS and Hi8 tape formats used in the prosumer systems, the imaging systems are much more sophisticated. Some of the best industrial video cameras may use up to three CCD units (one unit for each color of red, green, and blue) for impressive video quality. Resolution for industrial systems can exceed 450 horizontal lines. The industrial devices usually offer options such as industry-standard

Figure 10-6   Sony EVW-300 3-CCD Hi8 camcorder

SMTPE time code and image stabilization from time base correctors (TBC). SMPTE time code allows precise access to individual frames of video. Time base correctors provide general image stability under extreme circumstances. The prices for industrial systems start around $3,500.

Industrial-grade *video decks* are engineered to stand up to the demands of studio level work. When used properly and maintained regularly, these decks can accommodate 10 to 12 hours of constant daily use for years. Support for various forms of time code, such as industry-standard SMPTE time code, is usually built into these units. Industrial decks are only worth the additional investment if high-quality, hardware-based digital video capture components are to be used. Prices range from $5,000 to about $8,000.

tip

Always record video in SP (standard play) mode regardless of tape format or equipment quality. SP mode will allow two hours of video on a single VHS, S-VHS, 8mm, or Hi8 tape. Consumer and prosumer devices often provide additional support for EP (extended play) and SLP (super long play) modes. Using EP or SLP to record video will dramatically reduce video quality. Most industrial and broadcast video decks only support SP mode.

Broadcast

Broadcast-quality *cameras* offer the highest-quality video available in analog recording. They use triple CCD imaging systems and support the top-level tape formats such as 3/4-inch, 1-inch, and Betacam SP. Resolution for these devices exceeds 700 lines, surpassing the 640-line limit for digital video capture. The camera systems are expensive, ranging from $10,000 to more than $65,000.

With support for half a dozen formats and quality well above industrial equipment, broadcast quality *video decks* are a favorite choice of video professionals who can edit digital video offline for output back to videotape. Prices for new decks start at about $8,000 and can be as high as $65,000. Broadcast-quality decks add substantial performance gains to a digital editing system when compared to industrial-grade equipment. Broadcast-quality video equipment also offers substantial increase in quality over industrial devices when developing digital video that will be played back at full screen. But, as explained earlier in the chapter, broadcast-quality video decks provide only a marginal increase in quality over industrial decks when creating 320×240 digital video.

Shooting the Video

When shooting video footage that will later be digitized for use in a multimedia presentation, you need to consider several factors. The overall quality of any one step in the process affects all others, both positively and negatively. The disciplines involved here are not complex, but they do involve varying degrees of training and comprehension. Unless you intend that the final quality not have a polished look—when trying for an amateur, home-movie feel, for example—it might be a good idea to secure the services of a trained video professional.

Granted, using professionals can be costly and time consuming, two areas that digital video is supposed to reduce. Nevertheless, the time and money you might waste later when you are forced to dump the video for lack of quality could more than pay for a little professional guidance. You might consider, as many presentation producers have, taking a course in video production or hiring a video consultant to teach you the basics.

The following guidelines address a few of the common problems associated with shooting video. They will be useful when shooting video yourself or when giving direction to a video professional.

Not all video professionals are familiar with the demands of shooting for digital video. Before you make a hiring decision, find out if the person has shot video for multimedia production in the past. If not, be certain the person is aware of the basic rules that follow.

Crop the Shot

Because digital video is often played at sizes smaller than full screen, it may be difficult to see the subject in the video. Pay close attention to framing the shot so there is very little "extra" space around the subject. Figure 10-7 shows a good crop in which the subject is tightly framed and a bad crop in which the subject is diminished by the excess area in the scene. Most camera viewfinders are not accurate enough to provide a reliable framing guide. It is advisable to use a small liquid crystal camera monitor or a small video monitor whenever possible to ensure that the subject is framed properly. Color monitors will allow the videographer to check the light and color balance more effectively as well.

Figure 10-7　A tightly cropped subject (left) works better in a small window on the display screen

Shoot Slow and Steady Movements

If the video will play at less than 30 frames per second, excessive camera or subject movement will appear jittery or jerky. Since compression schemes are most efficient when adjacent frames are similar, excessive motion will reduce the quality of the video and increase its storage requirements. If the purpose of the video is to show an object in motion, care should be taken to limit the activity of the background while the subject is moving. Isolating background movement will also help focus attention on the subject. A tripod or stabilizer should be used to support the video camera whenever possible to avoid the influence of motion from the camera itself.

Use Direct and Even Lighting

Lighting the subject properly is just as important as limiting the motion. The video's subject should be evenly lit without being washed out or in shadow. Light sources can be used on multiple sides of the subject to reduce the effect of harsh shadows. The best lighting for digital conversion produces a soft transition from light to dark with no washed out or excessively dark areas. Natural daylight can provide a good source of light but can produce harsh shadows that do not translate well to digital video. Overcast or partly cloudy days are better because the sunlight is more diffused and will provide more even lighting.

Choose Subtle Color Schemes

Analog video information is separated into *luminance* (brightness) and *chrominance* (hue and saturation) information. Highly saturated and pure colors, such as red or blue, will reduce the overall quality of the video. Video can only carry so much information in the available bandwidth. Bright colors quickly use up the available bandwidth causing tonal variations to disappear. It is best to use subdued colors or a subtle range with only a few accents of bright color. Using subtle color allows more tonal information to be captured. The worst case for digital conversion would be a rapidly moving, bright red object.

Use Quality Audio

The most overlooked of the multimedia components is audio. Almost all video uses audio in some way. Use a professional-quality microphone only. Have the person appearing and talking in your video wear a microphone hooked directly to the audio-in jack on the camera (if available). If narration will be added to existing video, edit the audio and video together in the studio. Sound effects and audio from CD can also be added later in the editing process. If no audio is desired, remove the external microphone from the camera, turn off the built-in microphone on the camera, or insert a "dummy" plug into the audio in jack so that no audio is recorded.

Editing, Adding Effects, and Titling

Video footage can be edited either before or after it has been digitized. While this book is concerned with digital editing techniques for multimedia production, there may be cases when the video must first be pared down to a manageable length before digitizing—digitizing hours of footage requires huge amounts of storage capacity.

When videotape is edited and assembled onto a master tape, the process is referred to as *linear editing*. Videotape requires that each clip run in sequence from beginning to end. Even the addition or removal of a few seconds of material means a new master tape must be recorded. Finding the desired segments on a videotape is also a matter of advancing or rewinding to the right spot on the cassette.

Nonlinear digital editing, by contrast, involves first digitizing the videotape footage and storing it so that all points are randomly and almost instantly accessible to the editor. The clips are identified and marked, then rearranged and inserted in any order. When changes are made in the location, length of the clips, or transitions, the length of the entire program is automatically adjusted. Effects, titling, and sound tracks, too, can be added, altered, or deleted at any point in the timeline of the video.

Nonlinear Video Editing Systems

Roughly a half dozen companies are currently marketing dedicated nonlinear editing systems. They are many times faster than typical analog systems and cost less for comparable quality in most cases. Changes can be made with relative ease, allowing for last-minute additions or deletions that with an analog system would cost too much or seriously delay completion. This level of flexibility also extends the creative control of the producers and directors, allowing them to experiment more freely and letting them continue to shape the video footage until the last minute.

warning

The flexibility afforded by a nonlinear editing system can be a double-edged sword. Producers and directors must learn to resist the urge to tinker needlessly, wasting time and resources without seeing significant improvements in the final product.

New components and systems have dramatically reduced the cost of nonlinear editing in recent years—though systems are still expensive. As a result, many professional video producers and corporate video departments are using digital systems. Even though most of the systems are based on personal computer platforms, they are designed with proprietary software and are used primarily to edit analog video in a digital environment before outputting it back to videotape. Many of these high-end systems utilize elaborate multicard solutions to achieve superior image quality. They also use ultra-fast hard disks or disk arrays (multiple hard disk systems) to facilitate the high speeds of data storage, retrieval, and pass-through necessary to process video signals in real time.

Examples of nonlinear systems include the Avid Technology Media Suite Pro (see Figure 10-8), the ImMIX VideoCube, and the Data Translation Media 100. All three are based on the Macintosh operating system, though the VideoCube uses a proprietary platform running on a Macintosh. A few systems are available for the DOS and Windows platforms, including the Montage Group M series of products and PrimeTime Editor from Editing Machines Corp. Prices for integrated, nonlinear editing systems with add-on boards and software range from about $6,000 to $10,000. Standalone systems (with computer) cost $20,000 and up.

Digital Capture Hardware

After your video has been acquired on videotape, it is ready to be digitized. For this process your system uses a *video capture board*. Currently, there are more than 100 capture boards on the market, far too broad a selection to look at each product individually here.

Capture board prices range from less than $500 to a high of $6,000. For most presentations boards costing less than $1,000 will be sufficient. The Intel Smart Video Recorder for the PC, for example, lists for $699

Figure 10-8   Avid Media Suite Pro nonlinear editing system

and can be purchased for less than $500 in some markets. By using hardware-assisted and software playback, the board produces moderate-quality video in a reduced window size. The output works well for bringing video into a multimedia presentation that is not based entirely around the video segments.

If you are planning to experiment with video for a while before committing to a serious system, you might consider one of the ultra-low–price capture boards such as the Media Vision Pro MovieStudio (list price, $449). Boards in this price range give you basic video with few extras, but they are a good way to get started.

On the high end, boards such as the SuperMac Digital Film, RasterOps Editing Aces Suite, and Radius VideoVision Studio come close to the quality and performance of the integrated nonlinear systems. Based on the JPEG compression standard, these boards will capture up to 640×480 at a full 30 fps and 60 fields per second for maximum quality. Be aware, to deliver that kind of performance they must be paired with ultra-fast hard drives or disk arrays and powerful PCs. Prices range from about $4,000 to $6,000.

As explained earlier in this chapter, the final quality of the image is essentially a trade-off between window size, frame rate, and image quality (bit depth). When you look at the specifications for a board, remember that if its maker says the board can capture 640×480 (full screen) and 30 fps, it will not necessarily be able to do both at the same time. It may, for example, capture a full screen, but at a lesser quality; or it may capture a 30-fps image, but only at 160×120. Some boards that claim 640×480 do not capture at that size at all, but rather, capture at 320×240 and interpolate (double) the pixels upon playback. Boards that capture only 8 bits of information per pixel will not reproduce color well. A board of 16-bit or 24-bit capability is recommended.

When evaluating the capture capabilities of a board that claims 30 fps, check to see if the board is capturing a full 60 fields of information or only 30 fields. The latter represents only half of the information in a video signal. Finally, remember that a variety of variables can affect the final video, including signal noise, color shifts, timing errors, and artifacts from compression. Look at a few boards that fit your applications, your budget, and your system. Then rely on the subjective judgment of a professional product reviewer or your own eyes.

Quality Levels

The overall quality of video capture can be divided roughly into four levels: *basic or consumer*, *general business*, *professional business*, and *professional broadcast*. Each level requires greater skills, better tools, more time, and more money than the level before. Before investing in any one solution, it is advisable to carefully evaluate your intended application(s) and your commitment to using digital video.

Table 10-3 provides an overview of hardware. The products mentioned are representative of their class and are not necessarily meant as recommendations.

Digital Editing Software

Once the video is captured and stored on the computer, you may want to combine clips, add special effects, overlay animation or graphics, or add or remove audio. All these can be accomplished with digital *video editing software*. Editing software is often bundled with video capture cards for both the Macintosh and PC. Some editing packages offer special

	BASIC	GENERAL BUSINESS	PROFESSIONAL BUSINESS	PROFESSIONAL BROADCAST
Video capture (Sample hardware)	PC: Media Vision ProMovie Studio Mac: Built-in video compatibility on some models	PC: Intel Smart Video Recorder Mac: Supermac Video Spigot	PC: IBM ActionMedia II (two boards) Mac: New Video EyeQ	PC: None Mac: Radius VideoVision Studio
Window size	160×120	240×180	320×240	640×480
Frames/ fields per sec	15-24	15	30/30	30/60
Bit depth	16-bit	16-bit	24-bit	24-bit
Hard disk	230MB min. 400MB better 1GB best	230MB min. 400MB better 1GB best	400MB min. 1GB better 2GB best	1GB min. 2GB better 8GB best
RAM (minimum)	8MB–16MB	8MB–12MB	8MB–16MB	24MB–32MB
CPU (minimum)	486 /33MHz 68040/33MHz	486/66MHz 68040/33MHz	486/66MHz 68040/33MH	Power PC 68040/33MHz

Table 10-3   Video Capture Quality Levels

effects, but to maximize your effects capabilities you will need additional *special effects software.*

Entry-level digital video editors, such as those bundled with Quick-Time and Video for Windows, give you bare-bones editing capabilities. They will let you capture, play back and cut the video clips and may be all the editing power you will need for your video application. If you want to add transitions, special effects, graphics, or titles, you will need to step up to one of a dozen or so third-party products such as Adobe Premiere (used in the tutorial that follows), Avid VideoShop, ATI MediaMerge, or VideoFusion. Prices range from about $300 to $700 for most editing packages. Features, of course, vary with price.

Most of these editing programs import clips and represent them as small pictures (thumbnails) of the first frame. They typically use a timeline in which you select or drag a video clip and deposit it on the timeline in a desired sequence. The clips can be moved and adjusted at any point along the timeline, which can be set for a specific duration or set to adjust to the length of the audio and video clips it contains. The programs let you fine-tune the duration of the clips as well as the precision with which they connect. All editors include at least basic transitions and some limited special effects capabilities. In addition, several of the programs allow you to bring in special effects from other programs such as CoSA Effects, Digital Morph, Gryphon Morph, or MetaFlo. The high-end programs give you professional editing features such as advanced chroma key and compositing. Many can capture directly to RAM as well as to hard disk.

The best editing programs edit 60 fields per second—provided the video is captured at that rate—and give you quick previews of the movie you are assembling without the need to compile the movie first. Look for a program that lets you optimize your movie file for CD-ROM speeds (if that will be one of your delivery devices). A good built-in audio editor that allows you to adjust audio clips without leaving the program is also a useful feature. If you plan to invest in a high-end capture and compression system and will be editing for output to videotape, you will want to select an editing software package that supports SMPTE time code and will generate an edit decision list (EDL). These features, however, are not necessary for producing QuickTime and AVI movies destined to be part of a multimedia presentation.

Editing Tutorial

Regardless of the kind of digital video development you may be doing, and in most cases even regardless of platform, there are certain steps common to the process. This section will take you through the capture, editing, and compression process to get a feel for what is involved. Once you see the steps that are involved, you'll be able to better gauge whether you want to tackle this sort of thing yourself or hire someone else to help.

For this tutorial, the capture and editing were done on an IBM 486/33 with an Intel Smart Video Recorder, Adobe Premiere 1.1 editing software, Video for Windows 1.1, Microsoft DOS 6.2, and Microsoft Windows

3.1.1. This system provides good overall performance and is a cost effective solution for a variety of business presentation needs.

Capture and Editing: Adobe Premiere

Adobe Premiere is a full-function digital editing package available for Macintosh and IBM compatibles. While Adobe Premiere 1.1 has its own capture software, the tutorial uses VidCapture, a utility included with Microsoft Video for Windows. Adobe Premiere 1.0 does not include capture capabilities, but the version can be upgraded by calling Adobe.

Step One: Open the capture utility and select settings

The capture process begins by launching the VidCap utility from the Windows program manager to bring up the VidCap window. From within the utility, you have the option of capturing a single frame of video, multiple single frames, or motion video using the Capture menu:

Before choosing Video, use the Options menu to check that Video Format settings are correct:

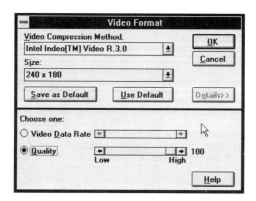

This is where you set your quality levels. The Intel Smart Video Recorder uses the new Intel Indeo Video R.3.0 format for best results. Here you also select the window size capture, in this case 240×180. There are also quality settings and data rate settings. For this example the quality is set to High.

For capture cards that support multiple video inputs, the Video Source selection under the Options menu lets you adjust image quality and compensate somewhat for poor lighting on a video by making a few adjustments:

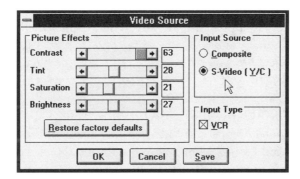

You can also set the source itself. If your capture card and video deck support it, use S-Video or Y/C input for best results.

Next you set the audio format. On the PC, the audio card usually works in conjunction with the video capture card to give you the combination of audio and video together. The Audio Format selection under the Options menu will give you limited control over your audio settings. It will allow you to set the sample size (8-bit or 16-bit), as well as the number of channels (mono or stereo) and the frequency (11KHz, 22KHz, 44KHz).

To start the capture, select Video from the Capture menu. The Capture Video Sequence dialog box lets you go to the audio and video options as described above in case you have not already done so:

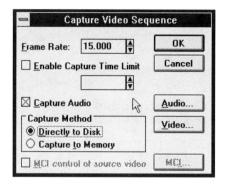

Set the frame rate for capture, the time limit, if any, whether you want to capture the audio, and your capture method (to hard disk or to RAM). When you select OK you are prompted to name the file. Once you do, the capture begins.

Step Two: Launch editor and import video clip

Open Adobe Premiere in Microsoft Windows. The screen displays four main windows—Construction, Transitions, Preview, and Info—where you will do most of your work (see Figure 10-9).

Bypassing Premiere's capture utility, you choose Import File from the File menu to reach a dialog box for locating the desired video clips:

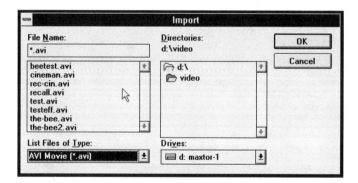

Premiere also offers the convenient feature of being able to import an entire directory of files at one time (Import Directory under the File menu). This is helpful when saving or organizing your files by project or by subject in directories on your hard drive.

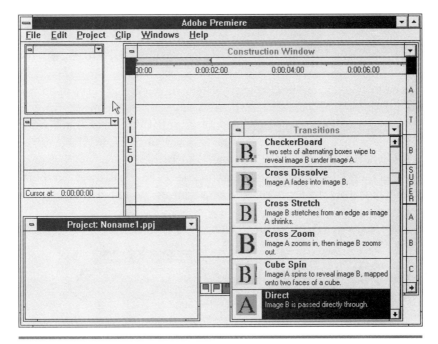

Figure 10-9 The main windows for Adobe Premiere

In Premiere, it is possible to import a wide variety of file types for images, audio, and movies. For images, PCX, BMP, and TIF files are most typical. For audio, WAV and AIF files are most common. To make your selections, double-click the file name in the list, or click once to select it and click the OK button.

Once you have imported all the clips and other files, Premiere lets you make notes in the area just to the right of each object in the Project window. These notes can be helpful in organizing the information to be edited, especially when more than one person is working on the project or when the small icons to the left of the objects are not sufficient to convey the true content.

Step Three: Edit the clip

To begin editing, select the clips in the Project window and then move them to the Construction window. Small preview images of the frames appear in the selected channel. In Premiere, you can manipulate two primary tracks of video (A & B). There is also a channel for transitions and one for superimposing graphics. More channels can be added as needed for more sophisticated projects. When a video clip is moved into place in the Construction window, any available audio for that clip is automatically placed in an audio channel as well. When you select a clip, the appropriate information also appears in the Info window (see Figure 10-10).

Once two clips are placed in the A and B video channels, trim extraneous information by using the Razor Blade tools available at the bottom of the Construction Window. Select where in time you want to cut and then select the unused areas and press the DELETE key. Graphics that are to be overlaid on top of the video can also be added to the SUPER channel just below the B channel:

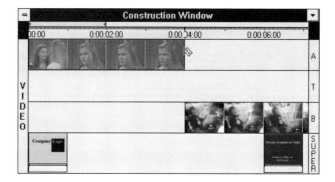

Step Four: Add transitions and graphics

To add one of the dozens of transitions, simply click on the desired transition in the Transitions window and drag it into place in the "T" channel in the Construction window. At this time you can also adjust the level of the various audio clips so that overlapping sections fade in and out smoothly.

Selecting any of the graphics to be used for overlay allows you to select the Transparency options from the Clip menu. When this option is selected, a dialog box appears and gives you a wide variety of control options over how that graphic is to interact with the video. (See Figure 10-11.)

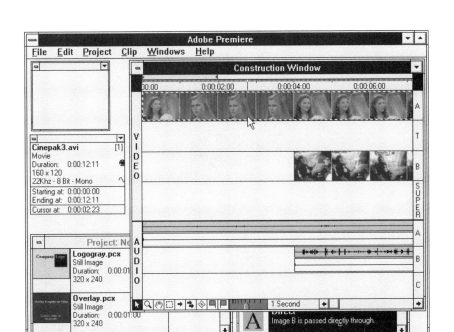

Figure 10-10   A video clip is selected and ready to be edited

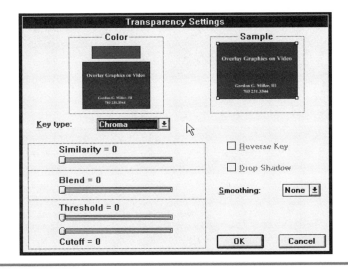

Figure 10-11   The Transparency Settings dialog box

Double-clicking an overlay graphic brings up a window with a full-size version of that image.

Notice the Duration button in the lower-left corner that can be used to set the exact length of time the graphic appears.

At any time, an area of the project can be viewed by selecting Preview from the Project menu.

Preview window

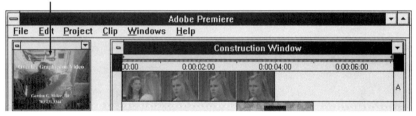

Notice that the preview in the Preview window is very small. This is done on purpose to allow faster processing of the information while editing. Adjustments can be made to the size of the Preview window by changing the Preferences for the program.

Simple editing controls are available at any time for video clips by double-clicking the frame in the Project window. Slider bars and directional controls let you mark in and out points. The same procedure can be accomplished with audio by double-clicking any of the audio clips.

Now that all of the edits have been made, there are a variety of choices to be made about the quality of the video clip and delivery environment. Before setting up the final compression settings, it is a good idea to save the project by choosing Save from the File menu.

Step Five: Compressing the movie

To create a movie, choose Make Movie from the Project menu. A dialog box will appear, asking you to give it a name and offering various options.

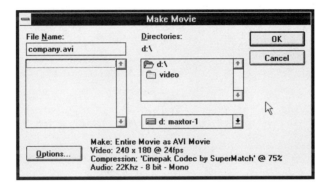

In this example, the movie will be output as an .AVI file. The window size for the movie has been previously set at 240×180; the frame rate is 24 fps. The Cinepak codec has been selected at a quality of 75%. Notice that the audio for this clip is set to only 22KHz - 8 bit - Mono. The relatively low quality setting for audio will allow the movie to run at a faster frame rate.

You can change any of the setting by clicking the Options button to bring up the Project Output Options window. There are options to set the data rate and delivery environment in this window as well. Once you have made all your changes, if any, click the OK button to go back to the Make Movie window. Give your movie a name and click OK.

The movie will begin compressing. Depending on settings and file size, this may take a while. When compression is done, a Clip window will appear with the final movie for viewing (see Figure 10-12). Changes can be made by going back into the project file, making adjustments, and outputting a new version.

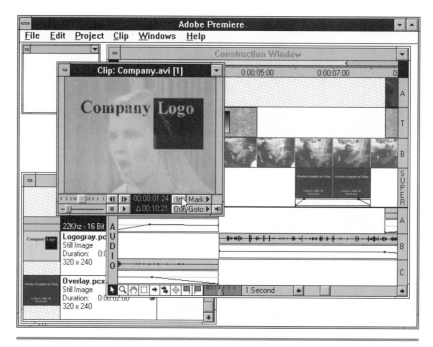

Figure 10-12 When compression is done, a Clip window appears with the final movie

Delivery and Display Options

Along with the decisions about capture and compression tools, it is necessary to determine your hard drive speed, hard drive capacity, and your RAM requirements. Remember, these will have an effect on the window size, frame rate, and image quality you will be able to achieve.

Hard Drives

Since most hardware-based digital video capture cards write information directly to the hard drive while digitizing, the hard drive performance is a major factor of digital video quality. Hard drives are manufactured by only a handful of vendors, but those vendors sell their drive mechanisms to dozens of other companies. You should not pay as much attention to brand names for hard drives as to their price, performance, and special features.

The specifications most frequently quoted are spindle speed, transfer rate, and access time:

Spindle speed refers to the speed at which the disk spins. Because of the volume of data recorded for digital video, a drive should have a minimum speed of 5,400 revolutions per minute (RPM). Some drives reach speeds of more than 7,200 RPM.

Transfer rate, discussed earlier, is the amount of information that can be transferred to and from the drive expressed in the number of megabytes per second. Retailers often boast transfer rates as high as 10MB/second. This figure can be deceptive. On the PC, the SCSI throughput will be a function of the quality of the SCSI card and the speed of the bus itself. Some fast PC SCSI cards can be limited by a slow computer bus. On the Mac, the built-in SCSI support moves at an effective maximum of 10MB/second so it is unlikely that drives will ever actually sustain such high rates.

Also, the quoted speed is often the maximum possible in *burst mode*, very short requests for information. When digitizing video, large streams of data are saved to the hard disk. Burst mode transfer rate is rarely a factor. Concentrate on the maximum sustainable data rate, which is often as low as 3–5MB/second. Some new drives can deliver a transfer rate greater than 10MB/second using ultra-fast SCSI-2 cards. These add-in cards, coupled with special software, increase the transfer rate by sending information across the computer's internal bus, which has a much higher sustained transfer rate.

tip

A drive quoted at a transfer rate of 6MB/second does not mean that you will be able to capture video at that rate. Some lab tests return drive results twice the rate of what you will get in actual operation. There are utilities available to verify the actual rate. If the product does not sustain the advertised transfer rate, take advantage of the manufacturer's 30-day guarantee by returning it.

Access time was once the standard by which all drives were judged. The specification refers to the time it takes for the drive to find a requested bit of information on the disk. These times vary dramatically from drive to drive and can vary with the method used to express the time factor. It is enough to say that if a drive has at least a 5,400 RPM spindle speed and a transfer rate of more than 5MB/second the access time is inconsequential. Just to have something to go by, times of 8 to 12 milliseconds are considered good.

Disk Arrays

These supercharged, hard drive speed demons use custom formatting software to make two identical high-capacity drives operate as if they were one drive. They write information to each disk in turn so that every other piece of information is found on each drive. This process allows for almost double the normal transfer rate of comparable single drives. Disk arrays are very expensive and only needed when full-screen, full-motion video is to be played back on the computer or output back to videotape.

In most cases, if full-screen, full-motion video is required, so is a disk array. If the digital video will be converted to software-only video at a size of 320×240 or smaller, a fast single drive should be good enough.

Drive Capacity

As mentioned earlier, you can never have too much hard disk space. This is most true when dealing with digital video. For example, assume that your high-quality digital video system is capturing video at full screen at a blazing fast 3.5MB/second. That is 210MB/minute. At that rate, a standard 2GB disk or disk array would hold a maximum of 10 minutes of video—editing, storing, and retrieving an hour of video would require more than 12GB of storage. For such extreme digital video needs, multidisk towers are available that store 28GB or more.

tip

The larger drives, in the 2GB range, are typically a better value as measured in cost per byte than smaller drives. Similarly, do not consider a multidisk array of less than 2GB if you want to get your money's worth.

Random Access Memory (RAM)

Back when digital video was becoming a viable option, most capture cards achieved better results when capturing to RAM. However, systems with as much as 64MB of RAM offered the ability to capture only 30 seconds of high-quality digital video.

As hard disks became faster, hardware developers moved away from RAM as the primary capture memory. With RAM costing more than $50 per megabyte and hard disk storage costing as little as $0.75 per megabyte, the transition was inevitable.

Nevertheless, there are good reasons to consider investing in as much RAM as your budget will allow. Virtual disks or *RAM disks* can improve the performance of the software applications that capture and edit digital video. Some manufacturers have created external drives that are packed with nothing but RAM. These *RAM drives* offer storage as high as 512MB with the lightning-fast performance of RAM. Access time is not in milliseconds (1/1,000,000 of a second) but in nanoseconds (1/1,000,000,000 of a second), offering an order of magnitude boost in performance. These RAM drives are very expensive. Their prices vary with the current market price of RAM.

CPU and Platform

Regardless of whether you use an Intel-based processor found in PCs or a Motorola processor found in Mac computers, the processor speed is a central determining performance factor for multimedia production. A 486/33MHz processor is the absolute minimum for PC-based capture when using the computer's CPU. The Motorola 68040/33MHz gives good overall performance for this kind of capture on a Mac.

Most low-cost video capture solutions (under $500) rely on the computer's internal processor to capture video. A slow processor will limit capture to small window sizes (160×120) and slower frame rates (15 fps).

Hardware-based digital video cards reduce the demand on the CPU by dedicating a separate processor to video. In hardware-based digital video, the primary processor is less important. Other components of the development system have a greater influence on hardware-based capture. The SCSI controller and quality of the SCSI hard drive will be more important than the amount of RAM or the speed of the system's CPU. In most cases the perception of overall system performance will improve with a faster processor. So while the speed of the CPU might not affect the quality of video capture in a hardware-based solution, a faster processor may make working with the system far more productive.

Some hardware-based digital video solutions can reduce the demand on the CPU by dedicating a separate processor to video. The CPU speed will also affect how quickly the user can move through information in an interactive presentation. Also, the more effects and transitions, the more demanding the load on the primary processor. So, while the processor speed can be lower for a hardware-based video presentation, it may not be the best solution for all components of your presentation.

Video Display and Delivery

The *video display resolution* of the computer monitor will dramatically affect the quality of not only the video but of all elements in a multimedia presentation. Most multimedia systems today support a minimum of 256 colors (8-bit) at the standard screen size of 640×480 pixels. Eight-bit color is usually sufficient for text, most graphics, and animation. Video is rarely effective at this limited resolution. Analog video requires at least 2 million colors to render properly. The only exceptions to this limitation are hardware-based systems designed to provide 24-bit video but that only work with 8-bit display solutions. The algorithms are optimized for displaying in 8-bit. Most notebook computers used for portable presentations only offer 8-bit display support. This is another reason for considering all aspects of the development and delivery display options before committing to a video system.

Most current Macintosh computers provide optional support for displaying 32,000 colors (16-bit) built in. 16-bit display support offers sufficient resolution to allow photographic-quality still images at full screen and good-quality digital video in window sizes up to 320×240. Video display support as high as 16.7 million colors (24-bit or 32-bit) is desirable. This level of support is often referred to as true color or high color support. Very few computers offer fast enough true color support to be viable.

The *display size* will also dramatically affect the quality of your video. The physical size of the display has little to do with the resolution available for displaying information. In the past, 13-inch, 14-inch, and some 15-inch monitors would support only a resolution of 640×480. The 16-inch and 17-inch monitors supported 832×624 or 800×600. The 19-inch and 20-inch monitors supported 1024×768 and 21-inch monitors supported 1280×1024. Today, it is common to find large displays supporting a range of resolutions and larger images. (For more tips on large displays see Chapter 11.)

Supporting display resolutions beyond 640×480 is often impossible when using hardware-based digital video. Because of the limited resolution of NTSC analog video, most hardware-based systems are built around display solutions of no more than 640×480. Presentation devices such as LCD panels are currently limited to this 640×480 resolution as well. The 640×480 size is universally accepted, but certain needs might be better addressed with larger displays.

Some hardware-based solutions will not work at all on large displays. Software-based digital video may run rough on a larger display because of the amount of information that must be redrawn. Display cards designed to accelerate large displays can effectively cure this problem but are an expensive fix for a minimal increase in quality. New technologies such as MPEG have just begun to liberate presenters from this 640×480 limitation by providing video support as high as full screen at 1024×768.

Digital Audio Support

Audio is an often-overlooked digital video component. No matter how good the video image quality may be, if the audio is poor the intended effect is ruined. Audio can also be an important factor in determining the overall data transfer rate. A minute of 16-bit digital audio sampled at 44.1kHz takes up almost 10MB of storage. The video component of a 320×240 software-based digital video clip that has been data-rate limited to 200K per second also takes up roughly 10MB per minute.

Because digital audio at its highest quality takes up as much space as video, it would be impossible to deliver the high-quality audio and the lowest-quality digital video from CD-ROM. Certain hardware-based digital video solutions such as DVI offer efficient schemes for including high-quality audio while still delivering good-quality video. In most cases,

8-bit audio sampled at either 11kHz or 22kHz is sufficient. In the case of nonlinear video editing systems, 16-bit digital audio is used to provide the highest-quality audio possible. The 120MB/minute required for high-quality JPEG video makes the 10MB/minute needed for audio seem insignificant. In these cases, extremely fast hard drives and not CD-ROM discs are required.

CD-ROM Drives

There are dozens of CD-ROM drives on the market. Most are very good and any will get the job done. However, most multimedia standards are designed around support for double-speed CD-ROM drives. These drives are called double speed (300K per second) because the data transfer rate is double that of first-generation CD-ROM drives (150K per second). Most double-speed drives are rated by access time. A good CD-ROM drive will have a minimum access time of no more than 300 milliseconds and data throughput of at least 300K per second. For development, purchase a drive with the shortest possible access time and the highest throughput. (For more information see Chapter 11.)

While these double-speed drives are a welcome relief from the stifling 150K-per-second drives, they are no match for the rapid triple speed and quad speed multidisk network drives. These drives have destroyed any reasonable stability developed from the large installed base of double-speed drives. For now, however, because of this installed base, double-speed CD-ROM drives will be the minimum standards.

Delivery Concerns

Delivering a multimedia presentation, particularly one that contains video, off-site can be a technologically hazardous affair. Even when you ask specifically what kind of equipment is available at the location, the information can easily be incorrect or misstated. Do not think for a minute that just because the computer sitting on someone else's desk or in a conference room somewhere is identical to yours that your program will automatically run on it. Custom software configurations, custom networking, and external drive configurations are just a few of the factors that can complicate using a strange delivery station.

When dealing with video, first identify what system and digital video software version the off-site machine is using. Most software revisions are backward compatible. This means the new version of the video software will play all video created with older versions. The opposite is not necessarily true. For example, if your presentation is designed to support digital video under system 7.1.1 on the Macintosh using QuickTime 1.6.1 and the off-site presentation machine is configured with system 7.0 running QuickTime 1.0, you are out of luck. If on the other hand your presentation used QuickTime 1.5 and the delivery machine has version 1.6.1 installed, you are set.

Microsoft Windows users may not find Video for Windows as compatible as QuickTime. Changes generally need to be made to the AUTO-EXEC.BAT files and the SYSTEM.INI files on the delivery machine in order to accommodate new version revisions. Often, jumpers on a video card may need to change and system interrupts (IRQ) need to be verified. Depending on the way the files are stored, it may be necessary to install new drivers from the original floppy disks.

This procedure can take several minutes even when you know exactly what you are doing and have all the disks on hand. Macintosh users only need to drop the new extension into the System Folder and restart. This convenience will hopefully find its way to the Windows environment soon. It can take as little as 15 seconds to reconfigure and reboot a Mac with a new version of QuickTime. That may be all the extra time you have when delivering an off-site presentation.

Video Backup Plan

No matter how carefully you plan, there will be a time when nothing works right. Consider an analog video safety net for video-based presentations. It will not take long to run through the presentation while recording to videotape. VHS is the most common video format in the world. You can probably find a VHS machine anywhere you need one. If your video capture system does not offer an RGB to NTSC scan converter, you should consider investing in one—they come in a variety of prices with a variety of features. It may seem a sacrilege to commit your interactive multimedia presentation to linear videotape with a meager 250 lines of resolution, but should your system fail, that tape could save the sale or save your job.

CHAPTER 11

Better one safe way than a hundred on which you cannot reckon.

—Aesop

CURTAIN UP: DISPLAYING AND DISTRIBUTING THE PRESENTATION

The last step on the multimedia ladder—the step you will take once your presentation has been designed and built—is, paradoxically, one of the issues you must consider first. *How* and *where* the presentation will be delivered will color and direct decisions made throughout the production process. Most importantly, it will influence the way you *store, transport,* and *distribute* the final presentation.

Delivering the Goods

A presentation can be given once or many times. It can be delivered by the same person or different people. In the case of multimedia sales brochures, advertisements, self-directed training programs, press releases, product updates, and the like, it can also be widely distributed as a self-running presentation. Because of the ease with which digital media can be altered and repurposed, it is conceivable that a core presentation might be delivered in all of these forms, in many venues and by many people. For that reason, delivery and distribution must be carefully preplanned—much more so than when using traditional forms of presentation. (The influence of presentation environments on content is discussed in Chapter 4.)

Imagine you are a sculptor commissioned to create a statue. You begin not by sketching or modeling, but by asking questions about the intended location and the audience: Will the piece be indoors or outdoors? How big is the space? What is the lighting? Who will see the statue and from what angle? How will the finished statue be transported and installed?

Those questions are not unlike the questions you must ask yourself before beginning your multimedia project. A few samples:

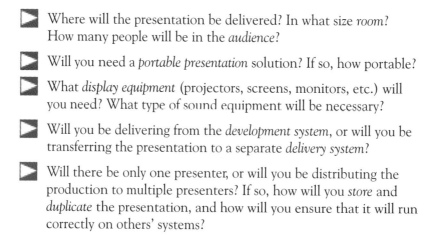

Where will the presentation be delivered? In what size *room*? How many people will be in the *audience*?

Will you need a *portable presentation* solution? If so, how portable?

What *display equipment* (projectors, screens, monitors, etc.) will you need? What type of sound equipment will be necessary?

Will you be delivering from the *development system*, or will you be transferring the presentation to a separate *delivery system*?

Will there be only one presenter, or will you be distributing the production to multiple presenters? If so, how will you *store* and *duplicate* the presentation, and how will you ensure that it will run correctly on others' systems?

There may not be a concrete answer to all of the questions before you start. But familiarity with the display options and distribution media will help you avoid costly and time-consuming miscalculations.

Evaluating Display Options

Regardless of the method of storage and distribution—both areas discussed later in the chapter—your multimedia presentation needs to make the audible and visual leap from the computer to the audience. This is the final link in the chain, and in many respects the most important. Your beautifully designed and executed multimedia project could degenerate into your electronic Waterloo if, at the moment of its presentation, the display technology proves inadequate or fails altogether.

The strategy you will use for evaluating and adopting display technologies can be summed up in one word: tradeoffs. There is no single solution that provides ideal results under even the most auspicious circumstances. The long list of variables begins with *cost* and goes on to include *portability, image size, resolution, brightness, noise, power consumption,* and *compatibility*. There are more, but that is enough for now. The message you should take with you is that even the very best display tools will run up against their limitations at some point.

As stated, your best defense against last-minute compromises and disappointments is to have the intended display technology in mind at the beginning of the project. Knowledge of *where, when,* and for *how many* will guide you through the process of narrowing the field of display options. Likewise, research into unavoidable details such as the availability of rental equipment and the necessary technical support for certain equipment will serve to short circuit possible disasters when the moment of truth arrives.

There are essentially five display types suitable for multimedia.

▷ CRT video/data projectors

▷ LCD projectors

▷ LCD panels

▷ Video/data presentation monitors

▷ Portable color computers

It is a good idea to become familiar with them all. Each has its range of prices, performance, and appropriate applications, as detailed in Table 11-1.

	CRT VIDEO/DATA PROJECTORS	LCD PROJECTORS	LCD DISPLAY PANELS	VIDEO/DATA PRESENTATION MONITORS	COLOR PORTABLE COMPUTERS
Image size	6' to 14'	6' to 10'	6' to 10'	20" to 40"	8" to 10"
Resolution	640×480 up to 1280×1024	640×480	640×480	640×480 up to 1280×1024	640×480
Cost	$10,000 to $50,000	$7,000 to $13,000	$5,000 to $9,000	$2,500 to $10,000	$2,000 to $10,000
Portability	No	Yes and no	Yes	No	Yes
Audience size	Mid to large	Mid to large	Up to 50	Up to 25	1 to 3
Environments	Auditoriums Training rooms Boardrooms	Training rooms Boardrooms Small auditoriums	Training rooms Boardrooms Small auditoriums	Boardrooms Training rooms Desktop / kiosk	Portable Desktop
Advantages	Stable Bright images Large images Ceiling mount	Convenient Large images Portable Ceiling mount	Light Portable Large images Convenient	Ambient light Sharp images Bright images Low maintenance	Very portable Convenient
Disadvantages	Expensive Bulky Needs dim light High maintenance	Low brightness Most 640×480 Needs dim light	Low brightness Requires OHP Noisy 640×480 Needs dim light	Heavy Bulky Image size limited to 40"	Small screens No audio Narrow viewing angle

Table 11-1 Multimedia-Capable Display Devices

CRT Video/Data Projectors

Cathode ray tube (CRT) projection units are the big guns of presentation display. They are most often used in training rooms, boardrooms, or auditoriums where they can effectively reach many people. Most can accept signals from computers or video sources, but have no onboard audio capability. For large installations, projectors are often hung from a ceiling or concealed in a pod at the front of the presentation room. They can also be placed inside elevation devices and be lowered from the ceiling or raised out of a cabinet.

The average units can provide large, sharp images up to 14 feet diagonally—specially equipped models can produce even larger images. Because the brightness drops off rapidly as the distance to the screen increases, the presentation room must be dark or very dim in order to effectively project and view a large CRT projection image. Highly focused task lighting may be required to provide lighting that does not interfere with the projection of the computer or video image.

Lighting problems can be partially overcome with a rear-projection system installed in a storage bay behind the projection wall. The image is projected on the back side of a special screen fitted to the wall. Rear projection systems offer a dark environment for the transfer of the image and allow the boardroom or auditorium to remain more brightly lit without dramatically affecting the overall quality of the projected image, though light that directly hits the screen will tend to wash out the image. The major drawbacks to rear projection are the extra space required, the lack of access to controls and connections, and the relative cost of building or refitting a room to accommodate a rear projection unit.

Prices for CRT projection units start at slightly more than $10,000 with some units costing well over $50,000. Most units are large and difficult to move. Built-in units, of course, are not portable at all. The typical CRT projector uses three CRT tubes, one for each color of red, green, and blue, focused through three lenses (see Figure 11-1). The lenses require calibration to keep them precisely aligned and, if the unit is moved, or even if the temperature in the room changes dramatically, recalibration is often necessary. While some newer models have a variety of automatic calibration systems, CRT projection systems typically require setup by a trained technician as well as regular maintenance.

warning

Proper color registration of three-lens projectors is crucial for quality images. Color shadows will make text unreadable and the image on screen will look like the pages from a 3D comic book—without the 3-D glasses.

LCD Projectors

Liquid crystal display (LCD) projectors are used in many of the same applications as CRT projectors, but bring with them a different set of trade-offs. Internally, LCD projectors use either a single-color LCD

Figure 11-1   Panasonic Model PT-200 large-screen CRT projector

screen or three (red, green, blue) elements along with a lamp to project a magnified image. Unlike most CRT units, most LCD projectors focus the image through a single lens, eliminating conversion problems (see Figure 11-2).

Because of fewer lenses and the lighter LCDs, these projectors are generally smaller, lighter, and therefore, more portable. The largest are as big as comparable CRT units, but the smallest are about the size of a thick briefcase. Most have built-in audio capability. Prices range from just under $7,000 to more than $13,000.

LCD projectors would seem to be the ideal multimedia display solution were it not for the limitations of the LCDs themselves. LCD elements are limited in resolution by the number of pixels they can fit into the area of the glass plates. Most can only project 640×480 or VGA resolution, a drawback for presentations that contain high-resolution images. Projectors also suffer from low light output. LCDs inherently absorb most of the light that passes through them—typically only about 2 to 3 percent gets through. The larger glass elements transmit more light,

Figure 11-2   Sharp XG-E800U LCD projector

but they are heavier and more expensive. The LCDs are also sensitive to heat, so a more powerful light source is not the solution.

The newest models have greatly improved the light transmission capabilities. Fast display elements have made it possible to display full-motion video. Continued improvements promise to bring performance closer to that of CRT; but for now, the trade-off between the two types of projection comes down to issues of price, portability, brightness, resolution, and ease of use.

LCD Projection Panels

LCD panels, which predate LCD projectors, are a hybrid projection solution. They convert data and video signals to an LCD display image, but require an overhead projector (OHP) for illumination and magnification. By separating the LCD display element from the projection components, the panels are made portable enough to carry in one hand (see Figure 11-3). But, they are forever dependent on the availability of an overhead projector. In the long run, smaller, lighter, and brighter LCD all-in-one projectors will very likely replace LCD panels.

Figure 11-3   In Focus PanelBook 525 LCD panel fits inside a
briefcase

Prices for active matrix LCD panels, models capable of displaying high-contrast, full-motion images, begin at about $5,000 and go up to nearly $9,000. The panels, like the overhead projectors they cannibalize, are useful for small- to medium-sized groups of up to 50 people or so. The images can be projected to roughly 10 feet diagonally; but as with other projectors, the brightness drops off quickly as distance increases, requiring the room to be darkened.

The panels suffer from the same low light transmission as the self-contained projection units. The brighter the projector—manufacturers recommend a minimum 3,000 lumens—the brighter the image. Again, it would seem that a super-bright OHP is the answer; but as the projection lamp gets brighter, it also gets hotter—and the panels are subject to performance glitches or failure if operating temperatures exceed recommended maximums.

warning

Reflective overhead projectors, those with the light source above the image, do not work with LCD panels. LCD panels require illumination from underneath.

Another tradeoff for the panel/OHP combination is its somewhat awkward presence at the center of the room where the lens and mirror of the OHP stick up like the head of an ostrich, blocking the view of anyone seated behind. On the plus side, a panel can be removed from the overhead projector whenever necessary, leaving the OHP available to other presenters who may want it for displaying standard overhead transparencies.

LCD panels support up to 16.7 million colors in the most expensive models. For displaying color images, animation, and video, 185,000 colors is about the minimum. Display will improve with 256,000 or 2 million colors, but the jump to 16.7 million colors may not be noticeable. The performance of a panel depends as much on the electronics it uses to convert the computer signals as it does on the color resolution it offers, so there are distinct differences in performance from one maker to the next.

The highest-end units also handle video, either via a built-in converter or an external adapter. Video adds upward of $700 to the cost of a panel, but for multimedia applications video is an obvious requirement. A few models also have audio pass-through controls. As with the self-contained projector versions, most LCD panels display a maximum 640×480 pixels. LCD glass element manufacturers are currently developing panels with 1024×768 pixels that will support workstation-level display resolutions.

tip

Some LCD panels offer back-lighting adapters that allow LCD projection panels to double as desktop display devices. At least one manufacturer has gone the other direction and offers a portable computer with a removable screen that can be used as a projection panel.

Video/Data Presentation Monitors

Any monitor large enough to handle a small- to medium-sized group could accurately be called a presentation monitor. These range from a useable viewing area of roughly 27 inches up to 40 inches (see Figure 11-4). Depending on screen size, presentation monitors can effectively serve groups of two or three up to 30 or so. In some situations, monitors are placed at the sides of a room about every ten rows, extending the useful viewing area as many rows back as necessary. Monitors are suitable for boardrooms, conference rooms, and small training rooms.

Presentation video/data monitors are based on CRT technology and, therefore, come with all the inherent benefits and drawbacks. They produce sharp, bright images that can be viewed in a room with normal illumination. Most support 640×480, 832×624, or 800×600 resolution; a few can handle 1280×1024; and the most expensive ($20,000) can support resolutions as high as 1600×1200.

The main drawback to these devices, other than their 40-inch image limit, is the huge glass picture tube. With tube and cabinet, the larger versions weigh nearly 300 pounds. Even mounted on a rolling cart, they

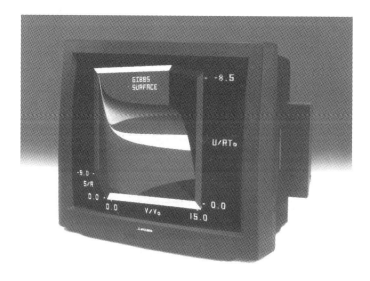

Figure 11-4   Mitsubishi AM4201R 40-inch presentation monitor

are difficult to move from room to room and are not viable as portable presentation displays. Prices vary widely from a $500, 27-inch consumer television set with standard video resolution to more than $10,000 for a high-resolution, 42-inch, multisync monitor. Average prices for data/video-capable monitors in the 27- to 32-inch range run from about $2,500 to $7,500.

Color Notebooks and Portables

For presentations on the go, you can choose from a variety of portable solutions that address the needs of multimedia. Products range in complexity and prices from simple, $2,000 color notebooks to fully integrated portable solutions that can exceed $10,000. The more advanced units contain a hardware-based digital video solution, digital audio card, integrated speakers and a CD-ROM drive. The systems can range from true notebook weight to as much as 30 pounds.

Subnotebooks are usually smaller than 8.5×11 inches and will fit neatly into a briefcase. Many offer 8-bit color video support on built-in LCD screens. Subnotebooks often suffer from small keyboards, small screens, few built-in options, small hard drives, and limited connectors; but a few have beefed-up 16-bit color, active matrix LCD displays with large 230MB+ hard drives, and RAM expandability to 32MB.

Subnotebooks, notebooks, and portables come with a variety of LCD screen sizes. Buy the largest size available or that you can afford. There is a dramatic difference between an 8½-inch diagonal and a 10½-inch diagonal screen.

By linking a subnotebook to a larger desktop computer or placing it in a special docking device, it is possible to expand the usefulness of the machines and take advantage of such features as greater hard disk space, networking, RGB-to-NTSC conversion, digital audio, and SCSI support.

At roughly twice the weight of subnotebooks, *standard-size notebooks* rigged for multimedia have the space and power to include fast, active matrix displays, built-in SCSI, digital audio, microphone, external display support, and printer and communications ports. Some of the newest models support 16-bit color, and offer hard drives up to 500MB and RAM expansion up to 32MB.

Portables pick up where notebooks leave off. This class of machine sacrifices light weight and small size in favor of incorporating the processor power, RAM, disk space, and floppies of a desktop computer as well as video, audio, CD-ROM, and even speaker components. Some models include hardware-based digital video systems such as DVI, as well as Ethernet capability.

tip

If you will be presenting directly from a portable's computer screen, you will need the fast response time of active matrix LCD technology to adequately display digital video.

Still larger portables are sometimes referred to as "*lunch boxes*" or "*luggables*." Weighing in at 23 pounds or more, these systems can serve double duty as full-featured desktop systems when not being taken on the road. The larger cabinets allow for expansion slots and a full array of connectors. A luggable system for multimedia presentation is capable of incorporating all the necessary features, including lots of RAM, a gigabyte or more of hard disk space, CD-ROM, digital audio, amplified stereo speakers, hardware-based digital video, external SCSI controller, on-board video support for 16.7 million colors, full-sized detachable keyboard, and more (see Figure 11-5). For the most part, these systems are sold as packages, custom-configured to order. It is important to note that these types of systems do not run on battery power.

Presentation Environments

Just as you should consider the type of driving you do when you buy a car, it is important to consider the presentation environment when deciding on a display device. Not every display solution will work in every environment. Similarly, each presentation environment will have characteristics that make it suitable for certain display solutions.

As with the display products themselves, presentation environments present you with a host of tradeoffs. You might, for example, prefer to use a data/video projector when you make a presentation at a client's office,

Figure 11-5   Dolch "luggable" multimedia presentation system

but arranging to rent or haul and set up a bulky projector at each location may prove totally impractical. Likewise, though your marketing presentation runs beautifully on your small, convenient color laptop, you would not want to ask a half-dozen important clients to hunch around the 9-inch screen.

Among the potential delivery environments a multimedia presenter is likely to encounter

▶ Boardrooms and conference rooms

▶ Training rooms

▶ Auditoriums

▶ Desktop

▶ Notebooks and portable presentations

▶ Kiosks

▶ Public venues (trade shows, conferences, etc.)

Designing for the Display

A few tips to help you coordinate the design elements in your presentation with the display system:

▶ Graphics software can adjust the color, contrast, and brightness of the image as well as create a custom pallet suited to the display device output.

▶ Computers can display a wider range of color intensity than NTSC video. For display on a video monitor, use an NTSC "safe" color palette. Use an image editor such as PhotoShop to put your colors in NTSC-safe range.

▶ Make your presentation design elements fill the computer screen so that no windows, icons, or other desktop interface elements can be seen by the audience.

▶ Whenever possible test colors, fonts, and drop shadows on the presentation display system before using them in the presentation. Serifs and italic fonts are risky choices.

▶ Before presenting, do not forget to adjust the brightness and contrast of the projector or monitor using sample screens from your program.

▶ Make sure buttons and text will appear within a "safe" area on the screen so they will be visible if the projection system does not display the border areas.

Boardrooms and Conference Rooms

During the past few years, as multimedia presentation has grown in popularity, presenters have been anxious to take advantage of multimedia for internal presentations and reports. These internal presentations often find their way into boardrooms and conference rooms. To be used effectively, multimedia display technology must be carefully selected to match not only the size of the room, its layout, and the anticipated audience size, but the style of presentation as well. For example, is the style formal, sophisticated, and methodical? Or, do presentations in the room tend to be informal, off-the-cuff, and last-minute affairs?

Testifying to the growing importance of electronic presentation, many boardrooms and conference rooms are being designed or renovated to accommodate ceiling-mounted projectors or rear projection units housed within one wall or a special cabinet (see Figure 11-6). While the projection devices, switchers, screens, and custom fixtures make this an expensive solution, it has the advantage of being clean, elegant, and sophisticated. The technology is not intrusive and the systems typically function well in a variety of lighting conditions. Video and computer signal switching can be routed from a variety of locations, and touch panels can provide a presenter with control over virtually all functions in the system and the room, including lighting, window coverings, sound levels, and communications.

The only limitations for equipping a conference or boardroom are time, money, and space. Installations may feature multiple rear screens and projectors, disappearing screens, or even computer or video displays built into the conference table itself for each individual. Behind the scenes signal routing, switching, and amplification can be adapted to literally any form of media playback. The systems can also be connected to the outside world via a variety of telecommunications links.

Figure 11-6   Hughes Communications executive conference
room. (courtesy of American Video
Communications, Los Alamitos, CA)

Training Rooms

Training room requirements are similar to those for large boardrooms or conference rooms, though cosmetics may not be as much of a concern. In some training facilities, each user has access to their own computer. Because the training rooms are often organized in a layout that is wider than it is deep, the central image may not need to be quite as large as for a boardroom of the same square footage (see Figure 11-7). More elaborate training rooms provide support for multiple large-screen rear-projection displays. Some feature switching and routing to display any computer screen in the room on the central display.

Auditoriums

Outfitting an auditorium for multimedia presentation capabilities is similar to outfitting a conference or boardroom, except that the scale is much larger. Because of issues of light, acoustics, and seating, auditorium

Figure 11-7   Amgen Training Facility. (courtesy of Advanced Media Design, Calabassas, CA)

installations should be accomplished only with the assistance of experienced professionals.

Individual display tools, such as LCD panels, will most likely be inappropriate for an auditorium. The projection and sound systems should be located in a control room where switching and routing provide for hooking up not only the installed components but external devices as well. A good auditorium system should, for example, allow a presenter to connect his or her own computer for multimedia playback rather than being forced to trust the built-in components. The facility should have the necessary patching and power connections available to all areas of the stage and podium. Support for some form of removable media and/or network access should also be provided.

For environments in which the auditorium is not equipped for multimedia presentation, a few companies offer complete, cart-based solutions that can be rolled in and hooked up to the room's standard audio system and projection device. The cart holds all the necessary playback devices and gives the presenter remote control and preview capability via color LCD displays. One switch powers on all the components. While this solution can be expensive, it offers advantages in setup, operation, and portability for auditorium presentations.

Desktop and Laptop

As multimedia presentations become more and more commonplace, the desktop is likely to emerge as the most frequently used delivery site. The desktop environment is intimate and immediate. A few people can gather comfortably around a screen as small as 14 inches (see Figure 11-8). Moving up to a 20-inch or 21-inch monitor can give you a large enough image for half a dozen people to view.

The two advantages to desktop delivery are 1) the presentation plays directly off the system on which it was created, and 2) the audience is free to select options, make suggestions, and ask questions. Using the same display for development and delivery can offer savings over using smaller monitors for development and using separate larger and more expensive monitors only for presentations. As noted in the previous section on portable presentations, when using an LCD screen for desktop and laptop presentation, keep in mind the narrow angle of view.

Figure 11-8   Toshiba color laptop computer presentation

Kiosks

A multimedia presentation in which the presenter is absent usually happens on a system called a kiosk. *Kiosks* are self-contained units that typically let the user direct the course of the presentation. Kiosks can include specially designed cabinets, as they do when built for trade shows or retail environments (see Figure 11-9), or they may be simply a computer located at a desk.

Kiosks can include any number of additional media playback devices and are often linked via phone lines or network to a central server. When designing a free-standing kiosk, make it attractive, durable, stable, and self-explanatory. In-house kiosks without specially designed fixtures should still be designed with the idea that the user will be operating the equipment alone. It should, therefore, be easy to use and as durable as possible.

The use of a touchscreen is critical in the design of many kiosk-based presentations. Touchscreens allow the audience to use the virtual on-screen buttons by simply touching the glass in the right spot. This is a

Figure 11-9   Informational kiosk (courtesy of North
Communications, Santa Monica, CA)

very seductive and user-friendly way to invite use and exploration of the
program. Touchscreens add only modestly to the cost of the entire system
and are well worth the money, particularly when the kiosk will be heavily
used in a public venue.

Case Study:

Company: Nabisco Foods Group, New York, N.Y.

Business: Packaged foods manufacturer

Objectives: Create a traffic-stopping interactive kiosk to provide information about company products, boost sales, solidify brand names, and collect customer inquiry data.

Trade shows are the three-ring circuses of business and commerce—vendors vie for attention from dazed attendees using everything from two-story videowalls and daringly dressed models to sports personalities and dare-devil rollerbladers. In the booth, overtaxed salespeople briefly chat with customers and frantically grab business cards, hoping their memories and a few scribbled notes will make sense of it all after it's over. All too familiar with convention chaos, Nabisco Foods Group created a traffic draw it calls the Nabisco Interactive Information System (NIIS) and rolled it out last year at the National Restaurant Association Show, the nation's largest food show.

Consisting of four 7-foot-tall, PC-based touchscreen multimedia kiosks,

the NIIS is designed to entice trade show attendees to stop and navigate through an interactive presentation about the company's range of products. It lures them into entering their name, title, and address by offering free recipes by mail and a chance to win a Panasonic TV/VCR. It also provides useful information about merchandising programs. The contact data, along with product preferences and interests, is collected for later use by the marketing and sales departments.

The kiosks, designed and built for Nabisco Foods by Einstein and Sandom, Inc. (EASI), a New York City–based multimedia marketing

and training company, are constructed around 486 IBM PC-compatible computers equipped with 16-bit SoundBlaster Pro sound boards and 20-inch SVGA touchscreen monitors.

As the attendees approach the booth and the kiosks, they see a welcoming animation and hear music. They are invited to touch the screen and begin the presentation and, after a brief introduction, are asked to swipe their trade show ID card through the integrated magnetic strip reader.

The main menu allows them to explore recipes for a variety of popular Nabisco food products by touching on the product of their choice.

Nabisco Foods Group

Each selection enables viewers to review recipes as well as merchandising information that Nabisco provides restaurateurs to support the brand. This information is delivered using 256-color, digitized video images, animated computer graphics, and 16-bit sound and music.

One of the key features of the system is its ability to collect market data. Using the Asymetrix Multimedia ToolBook authoring program and a special interpreter programmed in Turbo C++ language, the ID information is brought into the presentation where it is then saved to the PC's hard disk. "It's much easier and more accurate to take the information from the card than to have the users type it in," says J.G. Sandom, EASI's chief creative officer. The presentation includes a brief market research survey as well as the ability to monitor how long each attendee spent with the presentation and which sections they looked at most frequently. At the end of each day, the information from the hard disk is downloaded and sent directly to Nabisco for processing.

By collecting the information and downloading it each day, Nabisco's fulfillment house was able to respond very quickly to customer requests for information, says Robert Hiller, senior account executive of The Food Group, Nabisco's New York City–based advertising firm, which initiated the NIIS project. "When the food service proprietors got back to their establishments, the material they requested was waiting for them," says Hiller.

"It worked so well at the show that we're planning to re-do the program slightly so that we can give it to every one of our salespeople to put on their laptop computers," reports Hiller.

Storage, Transport and Distribution

There is an old Chinese proverb that says, "Fuel is not sold in a forest, nor fish on a lake." At some point, your presentation must travel from the development environment in which it was created to the presentation environment where the audience will see and hear it. Even if you bring the audience in house for a presentation, there will be issues of display, storage, and transport to work out in advance. The environment and the display devices will impact not only the design of your presentation but the hardware you choose to move it where it needs to go.

When transporting anything, you must first pack it. You can "crate" your presentation using any of several storage media. The media you choose will be dependent on such factors as how many copies you need, what you can afford, and what the delivery system can accommodate. Many of the transport and storage solutions can be used for both one-time presentations and widely distributed presentations, but their effectiveness in terms of cost and convenience varies dramatically. The media in the following sections is discussed in terms of capacity, quality, and price. All three elements should be considered before settling on one solution and, as mentioned at the start of the chapter, the considerations should be made before beginning production.

Videotape

Analog video tape may seem to have no place in a discussion of digital multimedia delivery formats. As was explained in Chapter 10, videotape is not interactive and is not designed for storing and accessing digital files. Nevertheless, there are tens of millions of VCRs in use, making video an ideal mass distribution medium for certain linear presentations.

If you are planning to distribute only on videotape, by definition you are creating a linear presentation and you do not need most of the tools and techniques covered in this book. You may choose to do your video editing and special effects in a digital environment (see Chapter 10) but if video is your sole delivery media, you are not creating an interactive multimedia presentation.

Nevertheless, video can be a useful and cost-effective way to give others, those without the means to see or run the multimedia program, some idea of what the original interactive presentation was like.

If you decide to "dump" a linear version of your presentation to video, keep these points in mind:

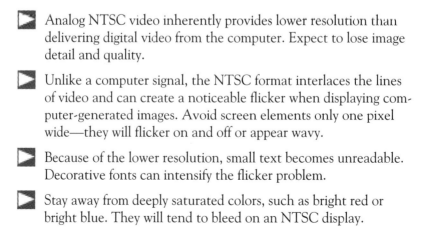

▶ Analog NTSC video inherently provides lower resolution than delivering digital video from the computer. Expect to lose image detail and quality.

▶ Unlike a computer signal, the NTSC format interlaces the lines of video and can create a noticeable flicker when displaying computer-generated images. Avoid screen elements only one pixel wide—they will flicker on and off or appear wavy.

▶ Because of the lower resolution, small text becomes unreadable. Decorative fonts can intensify the flicker problem.

▶ Stay away from deeply saturated colors, such as bright red or bright blue. They will tend to bleed on an NTSC display.

Conversion

Transferring your presentation to videotape requires that you convert the RGB computer output signal to NTSC video using a *scan converter*. These hardware components range in price from $200 to $20,000, depending on quality and features. They can be internal or external and fall into three performance categories.

Basic utility scan converters cost less than $500, but offer the lowest resolution and lack features such as *underscanning*, a technique that ensures that the entire computer screen image will be visible on the video screen. Some computers, such as the Macintosh AV models, have a good basic converter built in. Better models costing up to about $2,000 can do a significantly better job on resolution but still lack the extra controls for fine-tuning.

The next step up in quality starts at around $2,000. Most of these units provide good underscanning and include custom features such as image and color correction, convolution filters to reduce flickers and support for multiple video standards such as NTSC and PAL (the European television standard).

High-end scan converters start at $5,000. These devices are excessive in all but the most demanding cases where presentations are routinely transferred from the computer to video tape, or if the video will be broadcast. Converters of this quality should be used only with industrial or broadcast-quality analog video decks. Otherwise most of the benefits of the converter are wasted.

Floppy Disk

Almost every computer sold today has a built-in 3 1/2-inch floppy disk drive. The 3 1/2-inch drives have all but replaced the less durable and more limited 5 1/4-inch floppy disks. Even though floppy drives are more common than CD-ROM drives, it is nearly impossible (or extremely limiting) to run a multimedia presentation from a floppy disk because of its slow access time and limited storage. You can transport your multimedia presentation from the development station to the delivery station on floppies—watch out, you may end up carrying a dozen or more—or your presentation can be widely distributed on floppy disks to others.

In both cases, the production will need to be loaded from the floppies to a hard disk before it can be run effectively. Installation of multiple floppy disks can be difficult and time consuming. As a rule of thumb, you should limit the number of floppies for a distributed presentation to no more than three. If three disks will not hold the presentation with file compression, consider switching to a high-capacity distribution medium such as CD-ROM.

While effective electronic brochures and advertisements can and have been produced and distributed on floppy disks, the 1.4MB maximum capacity limits them to some nice graphics, a touch of sound, and a smattering of animation. There is very little room for full audio tracks and video is out of the question. File compression can yield as much as four times the physical limit of the disk. But beware, not all media files take well to compression.

tip

If you distribute a compressed version of your presentation, it is wise to make the files self-unpacking by integrating the decompression software on the disks.

CD-ROM

CD-ROM (Compact Disc-Read Only Memory) is the first cousin of the CD-audio format. In fact, the discs use the same physical materials and, in some cases, the same production techniques. Both are optical storage devices and both use lasers to read the data. The difference is that CD-ROM is designed as a storage medium for computer data, while CD-audio is formatted to optimize audio. Many CD-ROM players are able to play back CD-audio, but not vice versa.

The miracle of CD-ROMs is their ability to hold as much as 500 times the storage space of a single floppy disk. And, for widely distributed presentations—more than a 1,000 or so copies—the cost is roughly the same. (See Table 11-2.)

CD-ROM drives are quickly becoming a standard on Macintosh and Multimedia PCs. In 1993, there were 3.4 million drives sold, many as part of multimedia PC systems. In 1994, research firm Dataquest estimates manufacturers will sell more than 4.2 million units (about 80 percent will be used for multimedia applications, the rest for mass storage and retrieval). Predictions are that CD-ROM drives will soon become as

MEDIA TYPE	CAPACITY (ESTIMATED)	ACCESS TIME	MEDIA SIZE	MEDIA COST	DRIVE COST[1]	QUANTITY FOR 2.4GB	DISK+DRIVE FOR 2.4GB
Floppy disk	1.44MB	>1000 ms	3½-inch	$0.89	included	1,805 disks	$1,607
CD-ROM	650MB	<280 ms	5¼-inch	$1.30[2]	$0[3]	4 disks	$5.20
Optical[4]	128MB	32 ms	3½-inch	$34	$865	21 disks	$1,579
	650MB	28 ms	5¼-inch	$85	$2,169	4 disks	$2,509
	1.3GB	19 ms	5¼-inch	$112	$2,699	2 disks	$2,923
SyQuest	44MB	20 ms	5¼-inch	$59	$289	60 disks	$3,829
	88MB	20 ms	5¼-inch	$93	$429	30 disks	$3,219
	105MB	14.5 ms	3½-inch	$79	$375	25 disks	$2,350
	270MB	12.5 ms	3½-inch	$99	$699	10 disks	$1,689
Bernoulli	90MB	19 ms	5¼-inch	$105	$549	29 disks	$3,594
	150MB	19 ms	5¼-inch	$97	$649	18 disks	$2,395
Hard drive	270MB	12 ms	3½-inch	NA	$239	10 drives	$2,390
	540MB	9 ms	3½-inch	NA	$540	5 drives	$2,700
	1.2GB	10 ms	3½-inch	NA	$1,200	2 drives	$2,400
	2.4GB	10 ms	5¼-inch	NA	$1,800	1 drive	$1,800

[1] All prices subject to change.
[2] Cost assumes mass distribution media only and does not include cost of mastering and premastering.
[3] Cost assumes CD-ROM an integral component in a multimedia delivery machine.
[4] All media not standard on computers are priced as external SCSI devices. Internal devices are typically less.

Table 11-2   Distribution Media Comparison

standard as 3 1/2-inch floppy drives on new machines. This means that finding a presentation delivery system with a CD-ROM will soon be no problem at all. For wide distribution of multimedia presentations, CD-ROM provides a high-volume transport medium that can be played back on millions of machines.

Blank, recordable CD-ROM discs cost as little as $18 each and hold up to 650MB. When mass produced, the discs cost as little as $1.30 each. In addition to those benefits, CD-ROMs have an effective shelf life of more than 30 years. They are not affected by minor scratches or electro-magnetic fields, as are floppy disks, magnetic media, and hard disks. As a result, CD-ROMs offer an inexpensive, durable, high-capacity, and a widely available storage and delivery medium for multimedia presentations.

Formats

There are dozens of standards and formats for compact disc–based information storage. CD-ROM refers to the entire family of technologies represented by storing information on these discs. The basic standard for CD-ROM–based information, regardless of delivery format, is called ISO-9660. This is the standard sanctioned by the International Standards Organization (ISO). There are many file structure variations on this standard, designed to accommodate the variety of operating systems and file types. (See Table 11-3.) The structure you use will be determined by the platforms for both development and delivery as well as the nature of the information you are storing.

STANDARD	DESCRIPTION
ISO 9669 standard	Basis for all variations listed below. Has limitations in file structure, use of filenames, and number of nested directories.
HFS (Hierarchical File System)	Allows the familiar Macintosh graphical user interface to run on top of ISO-9660. Eliminates file structure, filename, and directories limitations.
Hybrid	Provides HFS and ISO support for Mac and PC using separate partitions.
Global hybrid	Offers universal access to both Mac and PC files on the same disk, but provides the HFS graphical user interface to Macintosh users.
Mixed mode	Allows one data track of either HFS or ISO-9660 information and subsequent tracks of CD-audio.

Table 11-3   ISO 9660 Standard Variations

Related Formats

The CD-ROM has been adapted to several other formats as well. Each is designed to optimize a particular type of content:

▶ *CD-I* (Compact Disc Interactive) was developed by N.V. Philips. Presentations authored on CD-I run only on CD-I machines. The machines output a video signal (NTSC or PAL) allowing the presentation to be played back on any television or video monitor.

▶ *Video-CD*, a relatively new format, uses the MPEG (Motion Picture Experts Group) compression standard to deliver up to 74 minutes of full-screen, full-motion video and CD-audio–quality sound from a CD-ROM player. MPEG adapters are available for some CD-I players.

▶ *Photo-CD* was developed by Kodak for both the consumer market and the business market as a means of storing and retrieving still images in a digital format—up to 100 images per disc. It is compatible with most CD-I players and with any multisession-capable CD-ROM drive running the appropriate software. Presentations created on Photo-CD using Kodak's software are limited to sound and still images.

Mastering

As the name implies, CD-ROM is read-only. Once information is written to the disc using a CD-ROM recorder (CD-R), it cannot be altered or removed. Two exceptions are the multisession CD-ROM recorders manufactured for CD-I and PhotoCD. These devices are capable of adding information at any time until the disc is full. They require a multisession CD-ROM drive to read the information added in subsequent sessions. Most new CD-ROM drives now have multisession capability.

Recording Hardware and Software

Do-it-yourself CD-ROM mastering is not for everyone. The prices for CD-ROM recording devices start at just less than $4,000 and range to $10,000 or more. Typical prices are in the $6,000 to $7,000 range. There are about a dozen models on the market. The most popular CD-Rs are made by Sony and Philips. The Kodak recorder, also made by Philips, is functionally identical to the Philips recorder.

CD-ROM recorders require special mastering software. While earlier versions of CD-ROM software were complex and difficult to use, the newer products are relatively easy to learn, making it feasible for you or someone in your organization to author CD-ROMs. Some manufacturers bundle software and hardware together, a good way to get started, but not always the way to be sure you are buying the best software.

Networked CD-ROM

Presentations, or parts of presentations, can be stored on CD-ROM and made available over a network. Network solutions can be effective for both development and for delivery (see "Networked Multimedia" at the end of this chapter). In the development process, the necessary production media can be stored centrally, allowing others on the net to access and use the materials. The same system can be used for storing completed presentations, which can be called up by others and run directly on their multimedia-capable terminals. Note that using this type of system for presenting is still a rare event and is fraught with technical issues, but will one day be common practice in every organization with the necessary multimedia network hardware and software.

CD-ROM storage on a network uses either a *multidisc* or *multidrive* CD-ROM server. Multidisc drives consist of a jukebox-like mechanism that feeds a disc into a single CD-ROM reader one at a time. Multidrive servers, on the other hand, have a separate reader for each disc, thus eliminating the delay time caused by the loading process. The advantage of networked CD-ROM over standard, high-capacity magnetic hard drives is one of storage costs. The price for a large-capacity hard drive is at least $0.75 per megabyte. The average price for a CD-ROM multidisc changer, by contrast, runs about $0.36 per megabyte, not counting the initial investment in the CD-ROM recorder. A 100-disc jukebox with 65GB of online storage reduces the cost per megabyte to as little as $0.18. Multidrive devices are a more costly option but provide instant access to all discs on the network without waiting for the changer to select one.

It should be remembered that CD-ROMs by definition are read-only. The information loaded in the server cannot be changed or updated without replacing or recording a new disc. Therefore, a CD-ROM–based server is best used for information that does not change frequently.

CD-ROM may be today's hot ticket for storing, presenting, and distributing presentations, but the technology is not without its shortcomings. In the first place, CD-ROMs are slow—hard drives are up to 20 times faster (see Table 11-2). Also, as mentioned earlier, information on CD-ROM cannot be easily changed or updated. For presentation situations that demand the flexibility to change data and require maximum speed and storage volume, some form of removable media or hard drive might be a better alternative.

Removable Media

Removable media consists of basically two types of technology: magnetic cartridge and magneto-optical. The first uses a magnetically sensitive disk in a cartridge that can be written to and read by a magnetic head the same way a hard disk functions. The second also uses changeable cartridges, but media is written and read with a laser. The physical setup for both is similar to CD-ROM; the drive is connected to the computer via a SCSI or other connector. The difference is that the removable media can be written to and from just like a floppy or hard drive. The CD-ROM, as explained, must be mastered with a CD-R and then played back from a CD-ROM reader.

Removable media allows the presentation to be stored, accessed, and changed without special mastering software or hardware. Separate cartridges can be used to store separate presentations, leaving the hard drive of the development system free for other work. Cartridges are available in a wide variety of prices and storage capacity. While their speed varies, they are typically faster than CD-ROM, and some are as fast as built-in hard drives. A carefully considered development and delivery strategy that includes removable media can eliminate problems associated with running out of storage space—when a cartridge fills to the brim, simply put in a new cartridge and continue working.

Removable media also can be a convenient way to transport a presentation from the development system to the delivery system, though cartridges are far costlier than CD-ROM. Removable media in the form of optical or magnetic disks can store from 44MB up to 1.3GB. With removable media, information can be written, erased, and rewritten many times. This ability to rapidly change, store, and update information makes removable media a good choice for demanding multimedia projects.

Removable media is not generally considered a viable option for widely distributed presentations due to the relatively small installed base of drives and the high cost of the individual cartridges.

There are numerous formats of removable media and a variety of storage capacities within each category. (See Table 11-2.) The right format and storage size depends on the demands of your presentation.

Magnetic Cartridges: Bernoulli and SyQuest

Bernoulli cartridges are based on the same magnetic storage technologies normally associated with floppy disks. Disks and drives are sold in three capacities: 45MB, 90MB, and 150MB. All drives are backward compatible—that is, a 90MB drive can use 90MB disks and 45MB disks but not 150MB disks, and 150MB drives can use disks of all three capacities. The Bernoulli disks have a more devout following on the PC platform than on the Macintosh, although drives exist for both. Until recently, Bernoulli disks and drives were much more expensive than the comparable SyQuest format. Early disk failures and high media cost have plagued this otherwise notable format. Price reductions have allowed Bernoulli to remain competitive in a market currently dominated by more stable and cheaper optical storage.

The SyQuest format is one of the oldest and least expensive forms of commercially available removable storage still on the market. SyQuest cartridges offer performance comparable to hard drives (see Table 11-2). The 44MB capacity was a major component of the desktop publishing revolution of a few years ago. With SyQuest cartridges, users of personal desktop publishing systems could send work to service bureaus and take advantage of services previously unavailable to them.

SyQuest continues to evolve its product line to address the needs of new technologies. Initially, there was concern about the format's susceptibility to damage by electromagnetic fields. While early users of the 44MB cartridges did report some problems, most were due to misuse or neglect. When handled and stored as directed, the cartridges provide years of reliable service. Recently, the company departed from its traditional 5 1/4-inch disks to offer a faster 3 1/2-inch, 105MB cartridge. The success of the 3 1/2-inch prompted the development of a new 270MB format that is currently one of the fastest forms of removable media on the market (see Figure 11-10). With access times and transfer rates

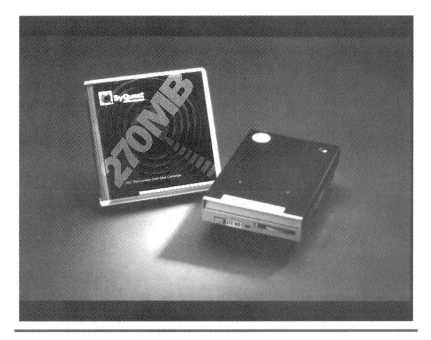

Figure 11-10   SyQuest 270MB drive

comparable to some hard drives, the 270MB cartridges are an excellent choice for multimedia presentations.

Magneto Optical Disks

Magneto optical disks use a technology more closely related to CD-ROM technology than to hard disks or floppy disks. Information is written to and read from the disk with a laser. Nothing ever comes into contact with the surface of the disk. Because information is read by laser, the disks are more stable and provide a longer shelf life for storing information. Sony, for example, guarantees their optical disks against failure for 30 years. Since these disks are not magnetic, they rarely suffer damage from electromagnetic fields that could damage magnetic storage cartridges.

Optical drives have a definite storage capacity advantage over other forms of removable media. For example, they have as much as four times the storage of the largest SyQuest cartridge and up to eight times the storage of Bernoulli drives (see Table 11-4). For large multimedia presentations containing large amounts of video or audio, optical may be the only viable form of removable media without going to a multidisk (cartridge) solution.

Optical disks have grown rapidly in popularity since the early 1990s. Their only serious drawback as a presentation delivery medium is their slow speed relative to other removable media (see Table 11-2). Even so, Pinnacle Micro (see Figure 11-11) already manufactures a drive with a rated access time of 19ms and transfer rates of almost 2MB per second, and rapid developments can be expected to continue to push the development of optical technologies closer and closer to the performance traditionally offered only by hard drives.

MAGNETIC DISK CAPACITIES			
SyQuest			
44MB	Single-sided	44MB/side	5¼-inch
88MB	Single-sided	88MB/side	5¼-inch
105MB	Single-sided	105MB/side	3½-inch
270MB	Single-sided	270MB/side	3½-inch
Bernoulli			
45MB	Single-sided	45MB/side	5¼-inch
90MB	Single-sided	90MB/side	5¼-inch
150MB	Single-sided	150MB/side	5¼-inch
MAGNETO OPTICAL CAPACITIES			
128MB	Single-sided	128MB/side	3½-inch
256MB	Single-sided	256MB/side	3½-inch
500MB	Double-sided	250MB/side	5¼-inch
650MB	Double-sided	325MB/side	5¼-inch
1000MB	Double-sided	500MB/side	5¼-inch
1.3GB	Double-sided	650MB/side	5¼-inch

Table 11-4 Removable Storage Media Capacities

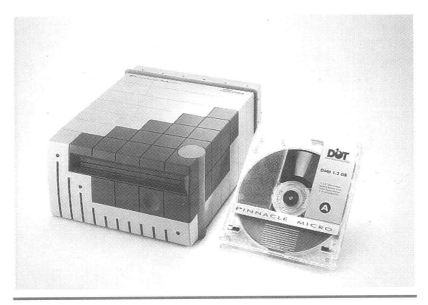

Figure 11-11   Pinnacle Micro optical disk drive

Hard Drives

Product for product, hard disk drives vary in storage capacity, speed, quality, features, and effective life, but to date they are unsurpassed for meeting the demands of multimedia presentation. In fact, in a presentation dominated by full-screen, full-motion video, hard drives may be your only option. (See Chapter 10 for a complete discussion of the role of hard drives in digital video production.)

Nevertheless, when compared with removable options, hard drives also have their drawbacks. External hard drives can be picked up and moved from place to place and machine to machine, but most are not designed for rugged transport. Connecting to a new machine can be both a nuisance and a liability—compatibility is never a certainty. Eventually, too, a drive will run out of storage space and the only alternative will be to dump some of the data or buy a new drive. The cost of buying a new hard drive is far higher than the cost of buying a cartridge with equal capacity.

A few manufacturers offer what they consider a best-of-both-worlds solution: *removable hard disks*. These are essentially external hard drives that have been configured to move easily from machine to machine via special docking ports. The ports are installed on the development and delivery stations; the drive slips out of one and into another. These removable hard disks offer the flexibility traditionally reserved for optical disks while maintaining the performance required for high-end, video-based multimedia. The major drawback lies in the cost of dual docking bays and, as mentioned, the cost of buying a new drive mechanism when adding an additional storage module. Removable hard disks offer up to 2GB of data storage.

Your best solution for presentation development will probably be a balanced strategy of using both a hard drive and removable storage. Use your hard disk storage for development and same-machine delivery. Use removable media for backup, storage, and transport.

Networked Multimedia

The Holy Grail of multimedia presentation production and delivery would be a broadband, wide area network that would allow any authorized employees on an as-needed basis from their own personal computers to access servers containing all relevant documents, figures, graphics, stills, audio, animations, and video. In addition, the network would allow the presenter to present to anyone, anywhere at any time—singly or in groups.

We are not there yet. But we are coming closer every day. Advances in networking systems as well as the communications infrastructure they run on are steadily moving toward a time when multimedia, with its time-based elements and huge bandwidth requirements, will be able to run down the wires (or fiber or airwaves) as easily as voices pulse through copper phone lines.

The obstacles are many and the largest are audio and video. To play audio and video over a network, a constant stream of high-speed data must be maintained in order to protect the integrity of the speed and sequence of the media. As it is, most network protocols handle all network requests with equal priority.

On a typical network, some users may be sending e-mail, others may be backing up files to a server, and others may want to view clips of digital media stored on a remote server. Requests for access are processed in turn regardless of the nature of the task. Also, on a network, information is exchanged in small chunks of data called "packets." This allows for easy management of the data. These packets are passed back and forth by the network controls with no regard to how much bandwidth a task-in-progress might require.

In the example above, the text-only e-mail message requires very little bandwidth to execute. Backing up files from a server can place a heavy burden on the network, but it is not critical to deliver data at a constant rate. Of the tasks, only digital video and audio are dependent on a constant stream of data.

Serving technologies are expanding to address the problem of pushing the vast data associated with digital video over the network as fast as possible. These video serving technologies are receiving widespread recognition due to their potential impact on the next generation of consumer-based multimedia services, notably interactive television and video-on-demand. The research invested in developing distribution schemes for consumer services will undoubtedly trickle down to network-based multimedia technologies for business.

Right now, there are only a handful of realistic solutions for delivering multimedia and video over a network. The most common solutions to date are delivered by Starlight Networks, ProtoComm Corporation, Novell, Inc., and IBM. The IBM solution requires an expensive hardware-based DVI card. The card must be installed in each delivery machine served and is only available to IBM PCs or compatibles. Other systems, like the Starlight Networks StarWorks media server, can manage and deliver a variety of video formats to PCs running DOS or Windows, Macintosh Computers, and soon to Sun Microsystems workstations running Solaris, a version of UNIX.

Most media servers provide support for a maximum of 25 users. Some systems can sustain 50 users, but this is still a long way from a viable solution to support the thousands of potential users for this technology. Fiber optic technology promises to provide all the bandwidth users need. However, it is currently too expensive to consider fiber optics as a universal network medium. Bandwidth is always the primary limitation of network-based multimedia solutions.

Hard disks and *disk arrays* offer theoretical performance as high as 15MB per second, with 3MB per second typical for fast disk array drives. Full-screen, full-motion video requires 30MB per second (with no compression) to play effectively. Compression schemes such as JPEG (Joint Photographic Experts Group) can deliver full-screen, full-motion video at 3MB per second. This would be an acceptable rate if you want to play

Servers and Workgroups

Networks and Servers

Networks of local machines in the same room, on the same floor, or in the same building are called LANs (local-area networks). A network connecting distant sites is referred to as a WAN (wide-area network). The granddaddy example of a WAN is the Internet. The Net, as it is known, links millions of users worldwide. With access to the Internet, users in remote areas can access information on servers and super computers all over the world. NCSA (the National Center for Supercomputing Applications) at the University of Illinois has made the single largest contribution to network-based computing. NCSA provides a host of network tools and resources to the public to make the network an attractive delivery medium. Recently NCSA has released Mosaic, an Internet-based informa-

tion resource that links local servers to a "world-wide web" of computing and information servers.

Despite its advantages, though, the distribution media needing the most improvement is the network. While NCSA's Mosaic is an extraordinary step in the right direction, there are some inherent difficulties. Networks have a limited bandwidth. Bandwidth is the physical limitation on the amount of information that can be transferred effectively at any given time. Bandwidth varies dramatically with the number of users. With only one task, one user would have the maximum bandwidth available on a given system. But a network attracts many users. Adding just one user to the network substantially decreases the available bandwidth. As networks grow to comprise millions of users, only a fraction of computing power is available to each user.

Small local networks usually do not suffer from the heavy traffic of wide area networks, even if their users can access the Internet. Effective networking strategies can insulate the typical corporate network from unwanted or unnecessary users.

When connecting multiple sites into a large wide area network, care needs to be taken to reduce excessive or unwanted traffic. Isolated local systems are fast. However, as soon these systems are connected to form a wider network, the system slows down. Today's widely available network protocols and infrastructures offer the greatest resistance to resolving this dilemma.

Workgroups

Small local networks can take advantage of well-integrated workgroup solutions that enhance productivity while managing resources with a central server. The Apple Share system software built into Apple's System 7 for the Macintosh provides an integrated solution for accessing information over both LANs or WANs. Microsoft Windows for Workgroups features similar solutions for PC users running Windows. Microsoft even offers a solution for DOS users who want to access Windows-based workgroup resources. These simple, closed-loop solutions are the basic model of an effective network.

one video to one person at a time over a small, high-speed network. Realistically, a rate of 3MB per second is likely to be reduced because of other demands on the disk such as file access and system-related tasks. The maximum sustainable data rate is likely to be less than 1MB per second. In order to deliver video at this low data rate, video window sizes must be reduced, frame rates have to be cut back, and image quality may need to be reduced to meet the demands of serving video to multiple users.

The server is typically connected to the hard disk via a SCSI controller. The *bus architecture* of the server plays an important role in determining just how fast data can be accessed from the hard drive. Some manufacturers use a fast 32-bit SCSI-2 controller to dramatically improve drive access. ISA-based PCs only offer a 16-bit bus and therefore cannot take advantage of 32-bit SCSI solutions. An ISA machine may be a bad choice for a server. Even the 32-bit solutions offered by EISA, MCA (Microchannel Architecture), and NuBus (Macintosh) computers only provide a maximum of 30MB per second. While this sounds like a lot, NetWare-based super servers can deliver almost ten times that rate. The Silicon Graphics Challenger Series servers, for example, can deliver a whopping 1.2GB per second across the internal bus. However, even if the bus could deliver this much performance, hard drives are still delivering only 3MB to 15MB per second.

The Bottleneck

A system is only as fast as its weakest link. Currently the network is the weakest link. Regardless of network type, the network is still considerably slower than either the hard disk or the server's bus. Ethernet has an effective maximum capacity of 1.25MB per second, but actual performance is much less. IBM's Token Ring offers only a 2MB per second effective data rate, but benefits from software that gives priority to time-based media. While there are radical solutions that can deliver over 12.5MB per second, such as *FDDI* (fiber distributed data interface), they are very expensive. FDDI cards can cost as much as $2,000. This cost can be prohibitive for a widely distributed network.

FDDI may soon be replaced by *ATM* (Asynchronous Transfer Mode). ATM offers the advantage of using existing switching technologies, similar to telephone switching systems, while providing more bandwidth and the ability to concurrently handle data packets and data streams.

The whole idea behind serving video is to reduce the burden on the *playback machine*. Ideally, the speed of the processor or the available hard disk space would not be a consideration. In reality, the speed of the CPU plays a role in how fast files can be accessed and saved from the server.

Some media servers utilize multiple Ethernet cards to distribute concurrent requests for digital video across separate branches of the same network. These systems effectively utilize the lower bandwidth of Ethernet. The branch of the network that serves the client machine does not require the high data rates associated with systems like FDDI. Client machines need only receive and play one digital video file from a server at a time. Therefore, Ethernet's 1.25MB per second data rate is sufficient to play back digital video on the delivery machine.

CHAPTER 12

To pretend to know when you do not know is a disease.

—Lao-tzu

MULTIMEDIA 911: FINDING PRODUCTION ASSISTANCE

Just as having the knowledge of where to find information is often more important than knowing it yourself, having access to the right production help is often the most important aspect of the multimedia production process. Aid is to be found in a variety of places and from a variety of people. The key is knowing what help you need and where to look.

Who Ya Gonna Call?

For all its do-it-yourself promises, computer technology is not and never has been a replacement for skill, talent, or experience. It may have put spreadsheets in the hands of the numerically clueless, but it has not replaced the need for accounting experts who understand what the numbers mean. Likewise, the desktop publishing tools that make it possible for anyone with preschool drawing skills to lay out and print a newsletter have driven home the point that creating a clear, consistent, and harmonious design requires more than access to a few templates.

These same realities apply to multimedia production by a factor of ten. Effective multimedia is a synthesis of seemingly antithetical disciplines: technology and aesthetics, engineering and art. Making all the pieces come together demands patience, persistence, skill, talent, and the intelligence to know when to call the cavalry. The key to finding the right assistance rests in understanding your own strengths and weaknesses as a multimedia artist and producer and in knowing where to find the help you need.

Do You Need Help?

The type and degree of help you need depends heavily on the size and scope of the project. Not every multimedia presentation requires the production crew from *Jurassic Park*. At the same time, unless you are a multimedia da Vinci—a genius at sound, video, animation, and graphics—chances are you will need at least some assistance for all but the simplest projects.

Consider the following criteria as you begin to design a project.

Time

Do you have sufficient time to design, develop, and debug the presentation yourself? Answering the question requires more than glancing at your schedule. You need to determine not only that you have the time, but whether the time could be better spent. If the project is not deadline- or content-critical and you have a sincere interest in building your personal multimedia expertise, then the long hours trying to do it all yourself become an investment in your education. If, on the other hand, time is tight and the project has little margin for error, you are well advised not to gamble exclusively on your own capabilities.

Technique

Are there elements of the production that lie outside the scope of your skills or technological resources? Know the difference between what you can accomplish and the services available from professionals. The gap is often larger than you suspect. If you decide to contract out for media elements such as graphics, animation, video sequences, or an original musical score, remember that you will still require time and talent to integrate the elements into the final product.

Perspective

This is the most overlooked reason for seeking outside help. You gain more from others than just their time and talent. Others bring fresh sensibilities and ideas to a project. They can help you focus on what needs to be said and what multimedia techniques might say it best. You may be an expert on the subject matter of your presentation, but that does not necessarily make you an expert on the subtle art of using media to capture and hold an audience. Without a critical eye and an impartial assessment of content, even the most elaborate big-budget presentation can wind up as boring as your neighbor's Disney World vacation videos.

Yes, I Need Help

If you have decided not to climb the multimedia mountain solo, the next step is to review your resource options. These fall into two categories: (1) in-house staff and (2) outside freelancers and multimedia production companies. Your final solution may be to assemble a team drawn from one or the other or from some combination of the two.

In-House Assistance

Start your search for assistance in your own backyard. Spend some time getting to know who and what might be available. Familiarize yourself with the various creative and support resources.

Specialized Department

The ultimate in-house resource is a resident multimedia production department. Companies with the internal resources to create full-blown multimedia productions are rare, but are coming online rapidly. If you have access to such a resource, consider yourself blessed. But just because the people and tools exist, do not take it for granted that they can accomplish what you have in mind.

The skills and systems of people who call themselves multimedia producers—even in-house—vary as widely as the talents of those ubiquitous individuals who carry business cards that say "management consultant." Evaluate your in-house department the same way you would an outside service provider. Follow the steps outlined later in this chapter. If you are under no corporate mandate to use the in-house services and they do not stack up to the production services you can get on the outside, then go outside.

Resident Talent

Even if your company does not have a department dedicated to multimedia, you might have access to someone in your organization who has become the de facto multimedia guru and can be equally helpful. As with desktop publishing and other computer-based tasks, one individual frequently gets tapped as the resident expert, usually because he or she likes to dabble with computers and software at home. This person may be able to help you produce your presentation or, failing that, help you evaluate the resources at hand.

Seasoned Veteran

There also may be people in your company who have undertaken projects similar to yours on their own, without using the in-house graphics or audiovisual departments. Find them. There is no substitute for experience, particularly with multimedia production. Experienced individuals can help you skirt some of the most common pitfalls, or they may be able to recommend valuable outside resource providers. At the very least, they can provide a reality check for your ideas and expectations.

Project Managers

If the core of a project is being kept in-house, one of the best ways to avoid headaches is to find someone to act as the coordinator or project manager. This individual serves as the point person whose responsibility is to see that everyone else executes his or her duties. The project manager does not necessarily need previous multimedia experience, just relevant project management skills.

Someone who oversees your company's annual report, advertising, or trade show planning is already familiar with the dynamics of bids, budgets, deadlines, and production schedules. Talk with the person to make sure he or she is comfortable with the idea of tackling the new and unpredictable challenges posed by multimedia production. Keep in mind that the project manager, too, will be educating himself or herself while scrambling up the steep and slippery learning curve.

Media Elements

When it comes time to collect the necessary media elements for your production, do not reinvent the wheel. Scrounge around in-house to see what might be available. Sift through the archives, consult other departments, take stock of relevant graphics that are currently available. If all else fails, put out a general call for materials on the company e-mail, in a memo, or in the company newsletter. What you need may be closer than you think. Consider the following examples.

Your Company Logo

The company logo probably already exists in a computer-accessible digital format in someone's computer. Why bother with the time and expense of redigitizing and manipulating it?

Artwork

Are there illustrations on hand that might be appropriate? Even outdated artwork can be scanned into the computer, updated with new data, and given a fresh look with graphics software.

Still Photographs

Are there suitable photos that have been taken for a brochure or an advertisement? Even if the images in print are not quite right, the photographer likely has other images on file. You may find just what you need among the rejects.

Video Sequences

Perhaps the motion sequence you need currently exists in a corporate promotional film, in a television commercial, or among candid camcorder footage from last year's company picnic. Virtually all formats of film and video can be digitized for use in a multimedia presentation, so if you can find it, you can use it—assuming there are no rights conflicts.

Outside Help

If you have explored all possible sources for assistance within your organization and still come up short on the necessary talent and resources, your only course will be to use outside services.

Freelancers

If you feel comfortable managing and authoring the project yourself, you may need to turn to freelancers only for individual media elements. Because you will need maximum ease and compatibility when incorporating the elements, it is important to find artists, photographers, musicians, and videographers who understand the intricacies of creating media for electronic presentation.

A good place to start your search is with freelancers who have worked with you or your company in the past. The artist who illustrated the company's fall catalog or the photographer who shot the annual report will have the advantage of understanding your company's products, style, and culture. Look for illustrators and photographers who can provide artwork in computer-ready form. Many artists now create work entirely in the digital environment. Others still use conventional artists' tools and then scan the result into a digital format. In either case, you will want to make clear your technical requirements for submission—file format, optimum resolution, storage medium, and so on.

 Graphic artists, who can help design the look and feel of your presentation. Although multimedia programming experience is not necessary, graphic artists should be well versed in the particulars of designing for multimedia. Dimensions and resolution, interface

Work for Hire

of the law, and therefore entitled to exclusive ownership of copyright.

▶ It precludes the freelancer from claiming additional compensation in the future, such as residuals, royalties, or commissions.

▶ It asserts that your company's relationship with the freelancer is one of client and supplier, not employer and employee, thus exempting you from the thicket of benefits, insurance, taxes, and employment laws. Note that in the eyes of the U.S. government,

Unless you intend to share the ownership of your project with your contributors, you would be wise to retain all freelancers under a written, legally binding "work-for-hire" agreement. This protects you and your company in a number of ways:

▶ It establishes your company as the sole "author" in the eyes

a work-for-hire agreement does not automatically keep a freelancer freelance. If he or she works exclusively for you for a significant period of time, or if the work was performed primarily in your offices, the Internal Revenue Service may consider the person a de facto employee.

The intricacies of work for hire and the fact that laws vary from state to state dictate the need for expert legal advice, both for drafting your own work-for-hire agreements and for reviewing work-for-hire clauses in the standard contracts of established freelancers.

---

elements, and overall visuals all need to be consistently applied, often over dozens or hundreds of individual screens. A graphic artist should be able to read and understand a project usage profile (see "Creating a Usage Profile" later in this chapter) and be able to operate within the framework of a decision tree.

▶ *Programmers* with solid experience in multimedia production. Proceed with a bit of caution here. Some computer freelancers, eager to stay as busy as possible, will bill themselves as experts in whatever programs are currently in demand. When interviewing potential programmers, use the same criteria for technical ability as you would for a full-service production house (see "Shopping for Multimedia Services" later in this chapter).

▶ *Writers* who are not intimidated by the challenges of nonlinear and interactive scripts. Writers specializing in nonlinear narrative are rare. If one cannot be found, look for a person with experience in industrial films, corporate video, and other long-form corporate communications productions. Seek out a writer who has exhibited the ability to work creatively with the dynamics of time and the juxtaposition of different media types.

Full-Service Production Companies

If the resources you need cannot be found in-house and the legal and logistical challenges of assembling a team of freelancers seems daunting, your next option is to contract with a full-service multimedia production company. In fact, this may be your first choice if you already know that you need and want soup-to-nuts assistance.

A full-service production company offers these potential advantages:

▶ A production company can be expected to deliver a finished product to which you hold a legally clear title.

▶ All work performed will fall under work-for-hire statutes. If the project contains copyrighted work, such as archival video or photographs, all rights will be previously negotiated and cleared, and the production company will assume liability for any mistakes.

▶ Most important, all production details will be someone else's concern, freeing you to focus on matters of content, delivery, and quality.

A full-service production company offers these potential disadvantages:

▶ Just like any other business, a full-service house has overhead and expenses that have to be met. These costs will be passed on to you, the client, both in professional fees and mark-up on materials and services.

▶ Although your say-so remains the final word, there is a difference between being a client and being the boss. Rather than being able to shape the project bit by bit through ongoing contact, you will need to fit your feedback into specific submission-for-approval points during the development process. If you try to translate a hands-on management style into a continuous stream of demands, comments, and criticism, you may end up frustrating both yourself and the production company.

For the production company, every project has a primary and secondary agenda. The first goal is to please the client and make a profit. The second is to create a production that will help it win additional clients. Watch out for a producer who tries to create a "showcase production" at your expense. You may find the company trying to convince you of the need for elements that seem inappropriate or excessive. Unless you are intent on being cutting edge yourself and cost is no object, draw the line whenever you feel uncomfortable with any suggestions that push the limits of what has been tried and tested.

Multimedia Consultants

If you are torn between using freelancers or a full-service production company, you have one more recourse: a multimedia consultant. Just as consultants do in countless other business capacities, presentation media consultants can help you navigate the often twisted and thorny path of multimedia production. A good consultant will

Help you define and refine the objectives of your project (see "Defining Your Objectives" later in this chapter).

Make specific recommendations on hardware, software, and creative resources.

Evaluate your proposals and budgets.

Evaluate bids and project submissions.

Oversee quality control and deadline compliance.

Act as a sounding board when new problems or belated inspirations arise.

Can you justify the additional cost of a consultant? If peace of mind is important to you, if you want to avoid late nights worrying about snowballing budgets and diminishing expectations, a consultant can be a reliable and cost-effective service.

Case Study:

Company: Health Net, Inc., Woodland Hills, CA
Business: Health services provider
Objective: Create a core marketing presentation to serve as the foundation for multiple presentation and training applications.

Health Net, a Woodland Hills, CA, health services company, knew from its research that with digital media it is possible to design a presentation so that its basic elements can be used in a string of related projects, without a lot of repetitive time, effort, or cost. But the company also knew it lacked the in-house resources to pull it off. It needed help.

The corporate communications department turned to an outside consultant, Los Angeles–based Lawson Group, to help them plan, produce, and create a presentation that could meet their multiple-use goals. Lawson, in turn, brought in multimedia producer Cimarron International, based in Aurora, CO, to design and author the individual multimedia elements.

"People tend to be reactionary about creating presentations," says Phil Lawson, president of the Lawson Group. "They do one thing at a time. Sometimes a company needs someone who can make a business case for taking a consistent modular approach. If this is done by someone inside, it can be difficult because of turf wars. It can be easier for an outsider to come in and suggest a strategic plan."

Lawson worked with Health Net's corporate communications department and sales staff to determine the objectives of each of the presentations and develop the content. Authored in Macromedia Director on a Macintosh IIci, the marketing presentation contains text, graphics, animation, and QuickTime video. It consists of separate modules, all of which are complete packages unto themselves, allowing presenters the flexibility to select only those portions of the presentation that are applicable to their particular audience.

Elements from the presentation were then combined with new sequences to create an informational presentation for the City of Long Beach, CA. This presentation, however, was designed for the CD-I platform, as well as the Mac. According to Don Cohen, CEO of Cimarron, several steps had to be taken to repurpose the presentation for CD-I.

HEALTH NET®

- 🌳 INTRODUCTION
- 🌳 PRODUCTS & BENEFITS
- 🌳 PROVIDER NETWORK
- 🌳 MEDICAL MANAGEMENT
- 🌳 WELLNESS
- 🌳 THE HEALTH NET ADVANTAGE

Health Net, Inc.

The video portions of the presentaion had to be compressed and encoded differently for CD-I. Even something as simple as the fonts had to be carefully tested to determine which could be carried across the Mac and CD-I platforms, as well as ported to videotape. To take advantage of the CD-I platform's higher-quality audio, Cimarron added more narration and longer music segments to the presentation.

Because Cimarron knew up front that the initial marketing presentation was going to be used for additional purposes and delivered on different platforms, the staff could plan the design to make the transition from Mac to CD-I easier. "If you solve problems right up front, it's going to save a lot of money," says Cohen.

Next, the same core marketing presentation was used and embellished to create a prototype of an information kiosk. Created for stand-alone use by employees at Health Net's member companies, the presentation is designed to answer employee's questions about their health plan as well as connect to a database that will allow them to make changes to their plan. The prototype also incorporates a search database of primary care providers. Browsing through the database, users can get a synopsis of a particular doctor, complete with a photo, hobbies, and directions to the office. Enrollment capabilities are included as well.

One additional adaptation: The prototype was used as part of an internal presentation to sell Health Net executives on the kiosk concept.

Of course, your peace of mind will be contingent on your confidence in the consultant. For that reason, you must put as much effort into evaluating the consultant as you would evaluating a full-service production company (see "Shopping for Multimedia Services" later in this chapter).

Defining Your Objectives

No matter who you eventually recruit for your project, the responsibility for getting off to a good start remains yours alone. As in many sporting events, a good start is the best determinant of a good finish. Begin by taking the time to visualize and express, as clearly as possible, what you want to accomplish. If that sounds painfully obvious, consider the case of the marketing director of a Midwestern bank, who will remain nameless to protect the producer who told this story.

The marketing director decided he wanted a multimedia presentation brochure that would describe a new set of services to prospective customers. Under tight deadline, he hired an outside firm to develop the presentation and put it on a floppy disk for distribution. The producer, working with little interference or supervision, delivered the first cut of the project, but the clever use of graphics, animation, and a bit of music were far, far away from the Star Wars–esque production the marketing director had envisioned. Disappointed and outraged, he fired the dumbfounded production company, hired a new company, and promptly had the same experience again. Later, the two producers compared notes, and it became clear that the marketing director's inability to communicate his goals and expectations had doomed the project in both instances.

In fact, most multimedia professionals will attest that the most frequent cause of client dissatisfaction can be traced not to slipshod work, but to vague, unrealistic, or unexpressed expectations on the part of the client.

Putting It in Writing

You can avoid disappointment by avoiding confusion. Before you approach anyone concerning your project, have a firm grasp on your intentions. Put them in writing. If you cannot set them down on paper, you probably do not know what you want. Remember: You are not writing a legally binding document. It will not be reviewed by corporate counsel or used against you in small claims court (probably not). This document is the first step toward clear communication and a successful presentation. As Ernest Hemingway would advise: Write what you know. Modify your project objectives whenever the need arises—but make those modifications in writing as well.

Begin the sketching-out process by first coming to grips with the key delineations.

Budget

Just how much do you expect the project to cost? Do your company's cash flow and financial controls create a preferred approach to payment, such as purchase order lump sums rather than incremental invoices?

Time

Exactly how soon does everything need to be done? Is it possible to begin with a prototype or to produce a less ambitious first edition before committing to the final product? You might consider building the full presentation in stages, modifying the design as you go based on feedback you receive.

Technology

For in-house work, does your system have the chops to do all you intend it to do? You must consider both the development and the delivery process. (See Chapters 3 and 11.) If going outside for services, do you have some idea of the tools and techniques that will be required?

Creating a Usage Profile

Once you have determined your qualifying factors, it is time to do the vision thing. At this stage, you do not need to worry about specifying content—forget content for the time being. Think about the aspirations of the product. What are you trying to achieve? What tone, style, or attitude are you trying to convey?

A useful exercise is to set down your goals in a document known as a usage profile. Usually only a page or two long, a usage profile sets the aesthetic and communication goals for the project.

Not all production companies or presentation specialists will refer to it by the same name, but all will recommend some form of the same process.

The purpose of the usage profile is to provide a sense of the project, its objectives and intents, to the contractors and in-house service providers. The example of a typical usage profile in Figure 12-1 demonstrates how, in about a dozen sentences, it is possible to clearly describe a project

MEMO

From: Ted E. Bear
       Marketing and Sales Director
       Squeezums Toy Co.
       Whiffenpoof, Alabama

To: MegaMultimedia Productions

Re: Trade Show Display

In this year's International Toy Fair, our company's booth will be located away from the high-traffic areas dominated by the major brand-name corporations. We are, however, near one of the main conference rooms, and crowds of people will pass us on their way to seminar sessions.

I want a multimedia display that will stop many of these people in their tracks, or at least get them to detour into our area for a minute or two—long enough for a personal contact by one of our salespeople. The display should probably last about 5 to 7 minutes and be continuously recycling. I don't foresee the need for interactivity.

The impact will probably be primarily visual. For instance, we might use three or four TV screens of about 20 inches each. Convention regulations strictly limit the amount of sound that can be projected into common areas, so we cannot rely on music and narration for impact.

We have a number of nice color photographs of our plush-toy line, plus some good footage of our soft-foam gliders in action. I'd like to use both of these, plus some kind of taped message from our CEO. Our marketing strategy is to sell toys not to kids so much as to parents ("The Toys You Wish You Had When You Were A Kid"), so images of happy families are important.

Because we want to use this presentation not just at this convention but at a few others over the course of the year, we need to have the built-in capability of replacing some of the plush-toy photos as our product line changes. I would also like to be able to use some parts during speeches. If the presentation is successful, we may decide to commission a Spanish-language version, so it would be good if the original is configured to make text translation as easy as possible.

Ted E. Bear

Figure 12-1   Sample usage profile

so that producers, artists, managers, executives, and anyone else involved in the project can quickly grasp its tone and intent. Note that the language used should be evocative of mood and direction, rather than being couched in the neutral and ambiguous phraseology of business-speak.

Shopping for Multimedia Services

Assuming that you decide to turn to outside help, with a budget, timeframe, and usage profile in hand, you can proceed with the process of finding and evaluating multimedia service providers—a term that for the purposes of this discussion covers freelancers, consultants, and full-service production companies. Remember: when shopping for services, shyness is a liability.

Ask potential providers for the following:

▶ A *cost estimate in writing.* In the case of freelancers and consultants, this should be for their hourly rates. For a production company, make it clear that you expect a formal proposal document. Unlike older industries that have established bid formats and protocols, the bidding process for multimedia services is not at all standardized.

▶ A *relevant résumé.* Because few universities or colleges offer academic degrees or formal certification in multimedia, you will have to look to a résumé rather than a diploma for evidence of professional standing. Try to look beyond job titles. An executive producer or similarly titled individual may have had little or nothing to do with the actual creation of the sample projects, whereas someone billed as a production associate may have been the visionary and driving force in the productions. Ask specific questions about what each job entailed and the degree of involvement of each individual.

▶ A *demonstration of earlier work.* It is a rare multimedia work that can be credited solely to one person. Most are collaborations of one sort or another. As just mentioned, when you review sample projects, make sure you know exactly which people should get credit for which aspects of the projects. For example, verify the programmer's claim to have designed the project or the designer's claim to have done the programming.

Look for a company or individual that can supply you with working samples that at least approximate the type of project you are planning. Just because a service provider has created a snazzy multimedia kiosk presentation for someone's trade show booth does not necessarily mean the provider will be the best candidate to build you a laptop-based interactive sales presentation.

Finding Service Suppliers

Where to look for multimedia expertise? Some firms are beginning to show up under the "M"s in the Yellow Pages, but most companies and freelancers are still a bit up river from the mainstream. The multimedia services business is undergoing explosive growth—so explosive that many professionals have found themselves able to stay busy without promotional efforts beyond the printing of a business card. To ferret out the top talent, you will probably need to do a little digging.

Referrals

Another company that has done similar work, and one that is not your direct competitor, may be able to give you a referral. Call friends and associates at other companies in other industries. Ask them if they have ever produced or seen a project similar to yours. If you are looking for a trade show display, walk through the exhibit hall of a show for an entirely unrelated industry. The fact that you are not caught up in the subject matter will help you concentrate on style and quality. If you see a presentation you like, stop and ask about how it was created and who did the work.

Current Suppliers

If your company has a relationship with another kind of creative service supplier, such as an advertising agency or graphic design firm, ask the supplier for multimedia producer recommendations. Many such companies have professional relationships with other service providers or have formed multimedia departments of their own. Again, be cautious.

You do not want your project to be the supplier's first attempt at multimedia, no matter how confident they may sound.

Hardware and Software Sources

The dealer or vendor from whom you buy your computer graphics or audiovisual equipment may have recommendations. Some dealers and vendors already have referral arrangements with multimedia service providers. Still others may have their own production service businesses. IBM's multimedia products sales division, for example, in response to perceived customer demand, formed the IBM Creative Services Group in Atlanta.

A number of multimedia software companies—San-Francisco–based Macromedia, for example—maintain certified developer programs. These programs are an excellent source for services because, in most cases, the individuals and organizations they list in their databases are carefully screened and prequalified. Apple Computer also has a referral service for its certified developers.

Keep in mind that vendors will naturally sponsor and promote service providers who sell or use the vendors' particular hardware or software. If the referral comes from a manufacturer, find out if the company is committed to one set of tools or a specific platform. The IBM Creative Services Group, for example, makes a point of stating that it is not committed solely to IBM products and will develop and deliver services using whatever systems are most appropriate for the particular project.

Trade Associations and User Groups

The multimedia industry is just beginning to organize. At present, there is no central clearing house or standard publication for referral listings of freelancers, consultants, or production companies. This situation is beginning to change.

Groups such as the International Interactive Communications Society (IICS) are making efforts to coordinate the listings and referrals in their local chapters to provide a national or international service resource. For now, it is best to contact the local organization, if there is one. The size and organization of each chapter varies according to the local participants. For a listing of local chapters, contact the IICS at (503) 579-4427.

The Interactive Multimedia Association describes itself as a nonprofit, international trade organization with more than 250 members; but as promising as the name of the group may sound for those seeking service referrals, the majority of the IMA members are large-scale hardware and software manufacturers. You can buy a copy of the IMA member directory by calling (410) 626-1380.

The Apple Programmers and Developers Association (APDA) is a source for Apple developer information and tools, but can also lead you to services. Call Apple Developer Hotline at (408) 974-4897.

The National Computer Graphics Association is not multimedia specific, but many of its members are skilled in related disciplines. Call (703) 698-9600.

The San Francisco Multimedia Developers Group (SFMDG) has both freelancers and multimedia production companies in its ranks. Call (415) 553-2300.

Computer user groups are often an excellent local resource for multimedia production services. The groups also vary in size and professionalism. They tend to be geared toward computer use as a hobby rather than a profession, but most have special interest groups (SIGS) that focus specifically on multimedia. To find a user group near you, ask a local computer dealer or contact any of the larger regional groups, such as the Boston Computer Society, at (617) 290-5700; the New York Macintosh Users Group, at (212) 473-1600; the San Francisco Macintosh Users Group, at (415) 477-1868; and the Chicago Macintosh Users Group, at (312) 973-1145.

First Contact

Your initial contact with a multimedia production company will probably be informal, giving each of you a chance to get to know the other. At this first meeting, you should provide a copy of your usage profile. You should be prepared to answer specific questions. One of the questions you will likely hear is, "What's your budget." Do not feel pressured to divulge your budget limits at this early stage. Rather, be specific about your desired objectives and let the production company provide you with a range of prices based on varying levels of service and complexity.

It is considered common courtesy, and sound business practice, to notify the producer that you will be talking to more than one company. As with any negotiation, your objective should be to generate a certain amount of price competition among the service providers.

Beware of companies that seem to promise too much for too little. If you are tempted to go with the lowest bidder, first find out how and why they can offer high-quality services for rock-bottom prices. Maybe you will find that you are dealing with a group that is extremely efficient and has developed "templates" for productions of your type. Or perhaps you are dealing with a talented individual or individuals who have ultra-low overhead and love the work more than the money—these operations are gold mines, but do not last long. As mentioned before, you may also come across a company that is willing to work cheap in order to get its first project under its belt. Keeping the risks in mind, this may be a way to get the most for your money.

When evaluating production companies and individuals, judge them by the following criteria.

Technical Ability

Are their programming skills and hardware up to the tasks you are demanding? Some signs of slipshod work include inconsistent interface elements (for example, some onscreen buttons work, others don't), excessively long access times for video or animation, and crashes or other anomalies caused by such unanticipated behavior on the part of a user as mouse-clicking in an inactive area. If you see any of these problems in the sample work, say a polite "thank you" and walk away.

Creative Ability

Are their design skills what you would like them to be, both in terms of creativity and execution? Look for attention paid to the details that add up to a "designed" feel—consistent typefaces throughout, a cohesively applied palette of well-chosen colors, music and special effects that match the story line, and so on.

Business Ability

Are you comfortable with them as a business entity? Do they seem adequately staffed and well managed, or do they seem a chaotic, shoestring operation? Do they have an impressive client list? Do they provide references—names and numbers—without being asked? How long have they been in business?

Unfortunately, it is difficult to find a provider that scores well in all three categories. Some are technically driven, and others are little more than graphic design firms with a single programmer. Some operations have decidedly corporate trappings and charge accordingly, and others have no formal offices at all. No one way of running a multimedia business is inherently better than another. Before making a final decision, you must find a balance of qualities that meets your criteria and feels right.

The Proposal

As already noted, at this stage of the multimedia presentation game, there is no standardized proposal format. Some firms will draft a casual letter, and others will mount an in-office presentation—a multimedia production in itself. Whether you are presented with a spectacle or a simple letter, be alert for four specific elements to the proposal.

▶ A clear assessment of the project in plain language that demonstrates the company truly understands what you are trying to achieve.

▶ A description of the team that will be assigned to your project. It should describe all individuals, their talents and specialties, and should designate specific tasks to specific people wherever possible.

▶ A firm commitment to completing the project within your timeframe.

▶ A project cost estimate that not only sets a price but the terms and conditions of payment.

Some companies will provide a document of understanding (DOU) that delineates purpose, objectives, and assumptions (see Figure 12-2). A well-drafted DOU should provide a clear sense that the entire process has been thought through, from initial meeting to final presentation. The document should also spell out the specific expectations of conduct for both client and producer.

The Production Schedule

Once you have chosen a producer and either signed the proposal or drafted a letter of agreement, the next step will be to collaborate on a

DOCUMENT OF UNDERSTANDING

IBM Creative Services Group, Atlanta

Project
Customer Executive Presentation

Background
The client requires an interactive computer-based presentation that will allow him to give an overview of IBM's view on what multimedia is, how it is being used, and the technology, and addresses the worldwide revenue opportunity. This presentation has existed since 1992 and has been updated yearly.

Purpose
This project will provide an updated version of the existing customer presentation. The updated version will operate on newer technologies and incorporate current media-based information about IBM's products and opportunities in the media field.

Objectives
• Convert existing presentation.
• Make changes to the content as specified by the client.
• Prepare the client's equipment to run the presentation.
• Create a CD-ROM version of the presentation that can run on the client's equipment.

Assumptions
• The client has no new video to be added.
• IBM Creative Services will create no more that 91 new graphics.
• IBM Creative Services will make no more than the 25 changes that have been identified by the client.
• IBM Creative Services will create the artwork for CD-ROM.
• IBM Creative Services will convert the existing customer presentation and author the new graphics.
• IBM Creative Services will install operating systems on client's equipment.
• IBM Creative Services will test and send out for duplication the CD-ROM for the field.
• Any changes requested by the client will be an additional charge and will be estimated at that time.

Figure 12-2   Sample DOU (document of understanding)

Project Team and Responsibilities

The Project Team will consist of:

- Producer/Designer/Graphic Coordinator. Responsible for providing direction on the project and working with artist and author to provide design/development direction.
- Authoring Coordinator. Responsible for the installation of software/hardware on client's equipment and will oversee the coordination of the authoring.
- Graphic Artist. Responsible for the creation of new renderings as well as for making changes to existing ones.
- Author. Responsible for all program coding of this presentation.

Responsibilities of Client

- The client will provide all content by the specified date(s).
- The client will provide billing numbers so the vendor cost for duplication of CD-ROM for the field can be billed directly to the client.
- The client will provide all hardware needed to make this presentation run on the client's equipment.

Terms and Conditions

- The total time to complete this project is within the limits set by the client. Success depends on overall teamwork and meeting all deadlines.
- In exchange for services, client will pay IBM Creative Services $XX,XXX.XX.
- Services provided will not exceed XX hours during the dates of service.
- In addition, client will reimburse IBM Creative Services for actual travel and living expenses incurred in providing these services.

This Document of Understanding will terminate upon completion of this services engagement, which is defined as development of the Customer Executive Presentation, or upon the client terminating our services.

Client's Signature: _____

IBM Creative Services Signature:_____

Figure 12-2   Sample DOU (*continued*)

production schedule. Keep in mind that such a schedule represents a commitment on everyone's part. Your company must be prepared to submit materials and give approvals promptly and punctually. If you are not clear on the steps involved in each process, do not hesitate to ask questions. If the answers are uncomfortably technical, ask for a simpler explanation or ask for the technical steps to be put in writing and have a third party—perhaps a consultant—evaluate the process for you.

Once approved, production schedules should be viewed like airline flight schedules—if one plane fails to leave the ground, the next one can't land. Except when compliance means seriously jeopardizing the project to achieve punctuality, you should do everything in your power to stay on schedule. Multimedia by definition involves the coordination of dozens of different elements and can be ferociously unforgiving if a schedule goes awry. Time permitting, try to view the presentation at specific points in the production cycle—you are less likely to be surprised, shocked, disappointed, or horrified when you see the final result. Regular viewings will also help you determine whether the production is running on schedule.

If your project does fall behind, don't panic. Insist on a frank and detailed explanation for the delay from the producer. Make an equally frank assessment of the impact the delay will have on the success of the presentation. Remind the producer, if necessary, that the responsibility to meet the deadline is ultimately the producer's. If the producer is clear on your expectations and your intention to enforce the contract, it is likely that the company will find a way to get the work back on schedule, even if it means 18-hour work days and evaporating profit margins.

Evaluation Process

Assume a best-case scenario: The project has been completed on time, within budget, and with ample communication between client and producer. For the most part, the producer has stuck to the schedule, and you have given the producer ample feedback along the way. The project is presented to you. "Here it is. It's done," they say.

The first question you must ask yourself is, "Is it done?" Chances are the finished product will not be absolutely everything you had imagined. There is frequently a gap between expectations and reality. If you are disappointed, you must determine if your dissatisfaction is the result of overly high hopes, or if the company somehow failed to deliver.

Before you start lobbing verbal grenades at the producer, or before you call in the legal eagles to avenge you, follow these steps:

▷ Spend as much time experiencing the presentation as possible. Run the program for yourself as if you were seeing it for the first time. If it involves speaker support, go through the speech several times and take note of where you think there are weaknesses.

▷ Perform some "destructive testing." Intentionally try to make the presentation crash by clicking the mouse at the wrong time and in the wrong places. Enter erroneous information or interrupt the program in the middle of a sequence to see what happens. In other words, try your best to make the program bomb.

▷ Invite people who have not been involved in the development of the project to experience it as if they were the actual audience. Admit that you may be too close to the creative and content elements to have an unbiased opinion. Are all functions clear? Is it easy to use? Does it accomplish the communication objectives?

You will probably receive a number of suggestions for improvement— some you will want to incorporate immediately, some you will file away for future improvements, and some you will need to disregard for practical reasons. Before you sign off on the project, make certain you are satisfied. It may be helpful to consult the following checklist:

▷ Is the overall running time of the finished product correct?

▷ Does the presentation run well on the hardware configuration of your choice?

▷ Does the software occupy the appropriate amount of disk space?

▷ Can the presentation be operated by those who need to operate it in its real-world application?

▷ Does the presentation avoid crashing when the user takes incorrect or unnecessary actions?

▷ If the presentation does crash, does it do so in an "orderly" fashion, without permanently damaging hardware or software?

▶ Have the necessary legal clearances been obtained for all media elements?

▶ Does the product contain a valid copyright notice in the name of your company on the display or in the source code?

▶ If the ability for future modification was requested, has that ability been demonstrated to you?

▶ If necessary, has a run-time agreement been obtained for the authoring software?

Index

B

D

N

Q

R

S

ω

X

XGA and XGA-2 graphics standards, 241
Xircom case study, 150-151

Z

Zero amplitude, 288
Zooming, 229

TAKE THIS PAGE AND FAX IT

to begin receiving Robert L. Lindstrom's
MULTIMEDIA PRESENTATIONS REPORT

Rip it, slice it, snip it, put it in the fax and send it now
to receive your **FREE** sample issue of the
MULTIMEDIA PRESENTATIONS REPORT

- ◇ Stay on top of the trends in tools and techniques.
- ◇ Optimize your electronic presentation abilities.
- ◇ Leverage the time and money you spend presenting information.
- ◇ Open them door to greater prestige and productivity.

For a further description see the next page. ➤

☑ Yes. Please send me the next issue of Robert L. Lindstrom's **MULTIMEDIA PRESENTATIONS REPORT** – FREE! Yes, I said **FREE !!!**

FAX the following information to: 1-818-952-5262

_____
NAME

_____
JOB TITLE

_____
COMPANY

_____
ADDRESS

_____
CITY   STATE   ZIP CODE

_____
RETURN FAX # (important)   phone #

Please allow six to eight weeks for delivery of your FREE issue.
MULTIMEDIA PRESENTATIONS REPORT ◇ 2222 Foothill Blvd., Suite E-141, La Canada, CA 91011-11456

FAX

the other side of this page to receive a FREE copy of the

MULTIMEDIA PRESENTATIONS REPORT

"The combination of making concepts clearer, using high quality images, adding sound and motion with relative ease, making changes faster, and reducing production costs translates to more efficient communication and a more productive organization."

— Robert L. Lindstrom
Business Week Guide to
Multimedia Presentations

Receive monthly advice and analysis from experts in the fields of sales presentations, training, information management, and multimedia computing. Learn by example from case studies that follow leading-edge companies as they adopt and adapt digital multimedia for more effective presentations.

MULTIMEDIA PRESENTATIONS REPORT helps you understand what can be done then helps you figure out how to do it:

- ◇ Presentation Strategies
- ◇ Product Evaluation and Pricing
- ◇ Core Technology Trends
- ◇ Production Strategies
- ◇ Human Behaviors in Presentation
- ◇ Evaluating Multimedia ROI
- ◇ Intelligent Information Systems
- ◇ Multimedia Networking Solutions
- ◇ Electronic Meetings
- ◇ Workgroup Presentations

The MULTIMEDIA PRESENTATION REPORT is a complete and timely resource that is designed to help you increase productivity by harnessing the power of multimedia presentations. Use it as your personal guide as you develop your individual and corporate presentation strategy for today and tomorrow.

 Do not be left behind by the digital revolution that is transforming information presentation. FAX the other side of this page to receive your sample issue absolutely FREE.

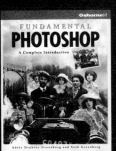

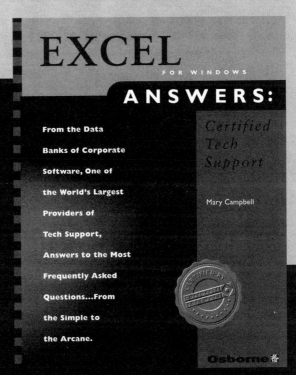

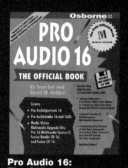

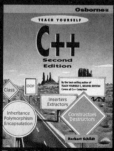

Secret Recipes

FOR THE SERIOUS CODE CHEF

BYTE's Mac Programmer's Cookbook
by Rob Terrell
Includes One 3.5-Inch Disk
$29.95 U.S.A., ISBN: 0-07-882062-6

No longer underground...the best-kept secrets and profound programming tips have been liberated! You'll find them all in the new BYTE Programmer's Cookbook series – the hottest hacks, facts, and tricks for veterans and rookies alike. These books are accompanied by a CD-ROM or disk packed with code from the books plus utilities and plenty of other software tools you'll relish.

BYTE's Windows Programmer's Cookbook
by L. John Ribar
Includes
One CD-ROM
$34.95 U.S.A.
ISBN: 0-07-882037-5

BYTE's DOS Programmer's Cookbook
by Craig Menefee,
Lenny Bailes, and
Nick Anis
Includes
One CD-ROM
$34.95 U.S.A.
ISBN: 0-07-882048-0

BYTE's OS/2 Programmer's Cookbook
by Kathy Ivens and
Bruce Hallberg
Includes
One CD-ROM
$34.95 U.S.A.
ISBN: 0-07-882039-1

BYTE Guide to CD-ROM
by Michael Nadeau
Includes
One CD-ROM
$39.95 U.S.A.
ISBN: 0-07-881982-2

BC640SL

LOTUS NOTES

ANSWERS:

Certified Tech Support

Polly Kornblith

From the Data
Banks of Corporate
Software, One of
the World's Largest
Providers of
Tech Support,
Answers to the Most
Frequently Asked
Questions...From
the Simple to
the Arcane.

Osborne

Lotus Notes Answers: Certified Tech Support
by Polly Russell Kornblith
$16.95 U.S.A.
ISBN: 0-07-882055-3

Think Fast
PASSING
LANE AHEAD

What's the quickest route to tech support? Osborne's new Certified Tech Support series. Developed in conjunction with Corporate Software Inc., one of the largest providers of tech support fielding more than 200,000 calls a month, Osborne delivers the most authoritative question and answer books available anywhere. Speed up your computing and stay in the lead with answers to the most frequently asked end-user questions—from the simple to the arcane. And watch for more books in the series.

**The Internet
Yellow Pages**
by Harley Hahn
and Rick Stout
$27.95 U.S.A.
ISBN: 0-07-882023-5

**Sound Blaster:
The Official Book,
Second Edition**
by Peter M. Ridge,
David Golden, Ivan Luk,
Scott Sindorf, and
Richard Heimlich
Includes One 3.5-Inch Disk
$34.95 U.S.A.
ISBN: 0-07-882000-6

**Osborne Windows
Programming Series**
by Herbert Schildt,
Chris H. Pappas, and
William H. Murray, III
**Vol. 1 - Programming
Fundamentals
$39.95 U.S.A.**
ISBN: 0-07-881990-3
**Vol. 2 - General
Purpose API Functions
$49.95 U.S.A.**
ISBN: 0-07-881991-1
**Vol. 3 - Special Purpose
API Functions
$49.95 U.S.A.**
ISBN: 0-07-881992-X

**The Microsoft Access
Handbook**
by Mary Campbell
$27.95 U.S.A.
ISBN: 0-07-882014-6

BC640SL

ORDER BOOKS DIRECTLY FROM OSBORNE/MC GRAW-HILL.

For a complete catalog of Osborne's books, call 510-549-6600 or write to us at 2600 Tenth Street, Berkeley, CA 94710

Call Toll-Free: *1-800-822-8158*
24 hours a day, 7 days a week
in U.S. and Canada

Mail this order form to:
McGraw-Hill, Inc.
Blue Ridge Summit, PA 17294-0840

Fax this order form to:
717-794-5291

EMAIL
7007.1531@COMPUSERVE.COM
COMPUSERVE GO MH

Ship to:

Name _____

Company _____

Address _____

City / State / Zip _____

Daytime Telephone: _____
(We'll contact you if there's a question about your order.)

ISBN #	BOOK TITLE	Quantity	Price	Total
0-07-88				
0-07-88				
0-07-88				
0-07-88				
0-07-88				
0-07088				
0-07-88				
0-07-88				
0-07-88				
0-07-88				
0-07-88				
0-07-88				
0-07-88				

Shipping & Handling Charge from Chart Below		
Subtotal		
Please Add Applicable State & Local Sales Tax		
TOTAL		

Shipping & Handling Charges

Order Amount	U.S.	Outside U.S.
Less than $15	$3.45	$5.25
$15.00 - $24.99	$3.95	$5.95
$25.00 - $49.99	$4.95	$6.95
$50.00 - and up	$5.95	$7.95

Occasionally we allow other selected companies to use our mailing list. If you would prefer that we not include you in these extra mailings, please check here: ☐

METHOD OF PAYMENT

☐ Check or money order enclosed (payable to Osborne/McGraw-Hill)

☐ AMERICAN EXPRESS   ☐ DISCOVER   ☐ MasterCard   ☐ VISA

Account No. ☐☐☐☐☐☐☐☐☐☐☐☐☐☐☐☐

Expiration Date _____

Signature _____

In a hurry? Call 1-800-822-8158 anytime, day or night, or visit your local bookstore.

Thank you for your order   Code BC640SL

About the CD-ROM

The CD-ROM that comes free with this book can be accessed by both PC and Macintosh users and includes:

▶ Free sample media clips that you can use with your presentation, drawing, or editing software.

▶ Demos and test drives of some of the most popular PC software applications for developing multimedia presentations, many of which are used in the tutorials throughout the book.

▶ Complete, installable versions of the PC programs that you can purchase by calling the listed toll free number to receive the necessary unlocking code.

▶ For Macintosh users, self-running demos and some working models of many of the products mentioned in the book and used as tutorials. (A disc with fully functional versions of the Macintosh products is available by mail to purchasers of the book.)

Caution: Some of the demos require 8MB of RAM. If your system has less RAM, you may not be able run some demos.

note

 A special Macintosh disc with full test drive functionality is also available. If you would like this free disc mailed to you, call InfoNow between the hours of 7 am and 9 pm Mountain Standard Time at 1-800-640-1853.

Some of the software you will find:

▶ Macromedia Action!

▶ Lotus Freelance Graphics

▶ Gold Disk Astound

▶ Fractal Design Painter

▶ Gryphon Morph

▶ Pixar Typestry

▶ Adobe Photoshop

▶ Adobe Premier

▶ Turtle Beach Wave for Windows

▶ Crystal Graphics Flying Fonts

▶ Free media clips and much more

For complete instructions on how to access and navigate the CD-ROM refer to the "About the CD-ROM" section after the Introduction to this book.

note

 If you have any technical questions with the CD-ROM, call InfoNow at 1-800-640-1853 and they will help you resolve your problems.

Disc Warranty

While we do everything we can to ensure the quality of this CD package, occasional problems may arise. If you experience problems with this CD, please call: InfoNow at 1-800-640-1853 (seven days a week, 7 A.M.-9 P.M. Mountain Standard Time).

WARNING: BEFORE OPENING THE DISC PACKAGE, CAREFULLY READ THE TERMS AND CONDITIONS OF THE FOLLOWING CD-ROM WARRANTY.

Limited Warranty

InfoNow warrants the physical compact disc enclosed herein to be free of defects in materials and workmanship for a period of sixty days from the purchase date. If your disc has a defect, call InfoNow at 1-800-640-1853 for a replacement.

The entire and exclusive liability and remedy for breach of this Limited Warranty shall be limited to replacement of defective disc and shall not include or extend to any claim for or right to cover any other damages, including but not limited to, loss of profit, data, or use of the software, or special, incidental, or consequential damages or other similar claims, even if Osborne/McGraw-Hill has been specifically advised of the possibility of such damages. In no event will Osborne/McGraw-Hill's liability for any damages to you or any other person ever exceed the lower of the suggested list price or actual price paid for the license to use the software, regardless of any form of the claim.

OSBORNE, A DIVISION OF McGRAW-HILL, INC., SPECIFICALLY DISCLAIMS ALL OTHER WARRANTIES, EXPRESS OR IMPLIED, INCLUDING BUT NOT LIMITED TO, ANY IMPLIED WARRANTY OF MERCHANTABILITY OR FITNESS FOR A PARTICULAR PURPOSE. Specifically, Osborne/McGraw-Hill makes no representation or warranty that the software is fit for any particular purpose, and any implied warranty of merchantability is limited to the sixty-day duration of the Limited Warranty covering the physical disc only (and not the software), and is otherwise expressly and specifically disclaimed.

This limited warranty gives you specific legal rights; you may have others which may vary from state to state. Some states do not allow the exclusion of incidental or consequential damages, or the limitation on how long an implied warranty lasts, so some of the above may not apply to you.

This agreement constitutes the entire agreement between the parties relating to use of the Product. The terms of any purchase order shall have no effect on the terms of this Agreement. Failure of McGraw-Hill to insist at any time on strict compliance with this Agreement shall not constitute a waiver of any rights under this Agreement. This Agreement shall be construed and governed in accordance with the laws of New York. If any provision of this Agreement is held to be contrary to law, that provision will be enforced to the maximum extent permissible and the remaining provisions will remain in force and effect.